Workshop on NLP for Medicine and Biology 2013

Associated with the 9th International Conference on Recent Advances in Natural Language Processing (RANLP 2013)

Hissar, Bulgaria
13 September 2013

ISBN: 978-1-62993-551-5

Printed from e-media with permission by:

Curran Associates, Inc.
57 Morehouse Lane
Red Hook, NY 12571

Some format issues inherent in the e-media version may also appear in this print version.

Printed by Curran Associates, Inc. (2014)

For permission requests, please contact the Association for Computational Linguistics
at the address below.

Association for Computational Linguistics
209 N. Eighth Street
Stroudsburg, Pennsylvania 18360

Phone: 1-570-476-8006
Fax: 1-570-476-0860

acl@aclweb.org

Additional copies of this publication are available from:

Curran Associates, Inc.
57 Morehouse Lane
Red Hook, NY 12571 USA
Phone: 845-758-0400
Fax: 845-758-2634
Email: curran@proceedings.com
Web: www.proceedings.com

Preface

Biomedical NLP deals with the processing of healthcare-related text—clinical documents created by physicians and other healthcare providers at the point of care, scientific publications in the areas of biology and medicine, and consumer healthcare text such as social media blogs. Recent years have seen dramatic changes in the types and amount of data available to researchers in this field. Where most research on publications in the past has dealt with the abstracts of journal articles, we now have access to the full texts of journal articles via PubMedCentral. Where research on clinical documents has been hampered by a lack of availability of data, we now have access to large bodies of data through the auspices of the Cincinnati Children's Hospital NLP Challenge, the i2b2 shared tasks (www.i2b2.org), the TREC Electronic Medical Records track, the US-funded Strategic Health Advanced Research Projects Area 4 (www.sharpn.org) and the Shared Annotated Resources (ShARe; https://sites.google.com/site/shareclefehealth/taskdescription; www.clinicalnlpannotations.org) project. Meanwhile, the number of abstracts in PubMed continues to grow exponentially. Text in the form of blogs created by patients discussing various healthcare topics has emerged as another data source, with a new perspective on healthrelated issues. Connecting the information from the three main sources in multiple languages to the scientific community, the healthcare provider, and the healthcare consumer presents new challenges.

The Natural language processing for medicine and biology workshop at RANLP 2013 provided a venue for presentations of current work in this field. The topics of papers and posters presented at the workshop included finding domainspecific symptoms in patient records, helping parents understand diseases, phenotyping, and deidentification of clinical text. We gratefully acknowledge the contributions of

- Sophia Ananiadou, University of Manchester, UK
- William A. Baumgartner Jr., University of Colorado School of Medicine, USA
- Svetla Boytcheva, American University in Bulgaria, BG
- Dina Demner-Fushman, US National Library of Medicine, USA
- Dmitriy Dligach, Childrens Hospital Boston and Harvard Medical School, USA
- Timothy Miller, Childrens Hospital Boston and Harvard Medical School, USA
- Sameer Pradhan, Childrens Hospital Boston and Harvard Medical School, USA
- Angus Roberts, University of Sheffield, UK

Organizers:

Guergana Savova (Children's Hospital Boston and Harvard Medical School)
Kevin Bretonnel Cohen (University of Colorado School of Medicine)
Galia Angelova (IICT Bulgarian Academy of Sciences)

Table of Contents

Workshop Programme

Friday September 13, 2013

9:00–9:15 Opening

Session 1

9:15–10:15 **Invited talk: Guergana Savova** (Harvard Medical School and Childrens Hospital Boston, USA) *Temporal Relations in the Clinical Domain and Apache cTAKES*

The presentation will consist of two parts. Part 1 will present an overview of methods and software development behind the Apache cTAKES platform (ctakes.apache.org). The second part of the presentation will shift to current research on temporal relations in the clinical domain. The research is done as a collaboration among Harvard, University of Colorado and Mayo Clinic.

10:15–10:45 *Finding Negative Symptoms of Schizophrenia in Patient Records*
Genevieve Gorrell, Angus Roberts, Richard Jackson and Robert Stewart

10:45–11:15 Coffee break

Session 2

11:15–11:45 *NLP can help parents to understand rare diseases*
Marina Sokolova, Ilya Ioshikhes, Hamid Poursepanj and Alex MacKenzie

11:45–12:15 *Active Learning for Phenotyping Tasks*
Dmitriy Dligach, Timothy Miller and Guergana Savova

12:15–12:35 *De-Identification of Clinical Free Text in Dutch with Limited Training Data: A Case Study*
Elyne Scheurwegs, Kim Luyckx, Filip Van der Schueren and Tim Van den Bulcke

Active Learning for Phenotyping Tasks

Dmitriy Dligach **Timothy A. Miller** **Guergana K. Savova**
Boston Children's Hospital and Harvard Medical School
firstname.lastname@childrens.harvard.edu

Abstract

Active learning is a popular research area in machine learning and general domain natural language processing (NLP) communities. However, its applications to the clinical domain have been studied very little and no work has been done on using active learning for phenotyping tasks. In this paper we experiment with a specific kind of active learning known as uncertainty sampling in the context of four phenotyping tasks. We demonstrate that it can lead to drastic reductions in the amount of manual labeling when compared to its passive counterpart.

1 Introduction

Several multi-year, multi-institutional translational science initiatives focus on combining large repositories of biological specimens and Electronic Health Records (EHR) data for high-throughput genetic research with the ultimate goal of transferring the knowledge to the point of care. Among them are Electronic Medical Records and Genomics (eMERGE)(McCarty et al., 2011), Pharmacogenomics Network (PGRN)(Long and Berg, 2011), Informatics for Integrating Biology and the Bedside (i2b2)(Kohane et al., 2012). In each of these initiatives there are a number of diseases or driving biology projects that are studied such as Rheumatoid Arthritis, Multiple Sclerosis, Inflammatory Bowel Disease, Autism Spectrum Disorder, Early Childhood Obesity, each defined as the phenotype of interest. To enable large cohort identification, phenotype-specific algorithms are developed, evaluated and run against multi-million-patient EHRs which are then matched against the biobanks for further genetic analysis. Efficient and accurate large-scale automated phenotyping is a key component of these efforts.

Supervised machine learning is widely used for phenotype cohort identification (Ananthakrishnan et al., 2012; Ananthakrishnan et al., 2013b; Ananthakrishnan et al., 2013a; Lin et al., 2012b; Lin et al., 2012a; Xia et al., 2012). However, the supervised learning approach is expensive due to the costs associated with gold standard creation. While large amounts of unlabeled data are available to the researchers in the form of EHRs, a significant manual effort is required to label them. In a typical phenotype creation project (Lin et al., 2012b; Lin et al., 2012a), a pool of patients is identified using some filtering criteria and a subset of patients is selected from that pool for subsequent expert annotation. During annotation, a domain expert examines the notes associated with a patient, assigning either the positive label (i.e. relevant for the given phenotype) or the negative one. A model is subsequently trained using the annotated data. This scenario is known as *passive learning*.

On the other hand, active learning (Settles, 2009; Olsson, 2009) is an efficient alternative to the traditionally used passive learning as it has the potential to reduce the amount of annotation that is required for training highly accurate machine learning models. Multiple studies have demonstrated that when active learning is used, machine learning models require significantly less training data and can still perform without any loss of accuracy. Active Learning is a popular research area in machine learning and general domain natural language processing (NLP) communities. However, its applications to the clinical domain have been little studied and no work has been done on using active learning for phenotyping tasks. In this paper we experiment with a specific kind of active learning known as *uncertainty sampling* in the context of four phenotyping tasks. We demonstrate that active learning can lead to drastic reductions in the amount of manual labeling without any loss of accuracy when compared to passive learning.

2 Background

2.1 Phenotyping as Document Classification

Phenotyping can be viewed as a document classification task in which a document consists of all EHR documents and other associated data (labs, ordered medications, etc.) for the given patient. Initial filtering is usually performed based on a set of inclusion and exclusion criteria (ICD-9 codes, CPT codes, laboratory results, medication orders). Within the eMERGE and PGRN, flowcharts outlining each phenotyping criterion and logical operators (AND, OR) are defined to constitute the phenotyping algorithm (Pacheco et al., 2009; Waudby et al., 2011; Kho et al., 2011; Kullo et al., 2010; Kho et al., 2012; Denny et al., 2010). The phenotyping within i2b2 takes a different approach – that of a machine learning patient-level classification task (Ananthakrishnan et al., 2012; Ananthakrishnan et al., 2013b; Ananthakrishnan et al., 2013a; Lin et al., 2012b; Lin et al., 2012a; Xia et al., 2012). Each patient is represented as a set of variables derived from the structured and unstructured part of the EHR (ICD-9 codes, lab results, relevant mentions in the clinical narrative along with their attributes) which are then passed to a machine learning algorithm. Whether the choice is a rule-based or machine learning approach, a fairly big sample of data needs to be labeled by experts which will then be used to derive the rules/train a classifier and to evaluate the performance.

2.2 Active Learning

Active learning is an approach to selecting unlabeled data for annotation that can potentially lead to large reductions in the amount of manual labeling that is necessary for training an accurate classifier. Unlike passive learning, where the data is sampled for annotation randomly, active learning delegates the data selection to the classifier. Active learning succeeds if it reaches the same performance as its passive counterpart but with fewer training examples.

Seung et. al. (Seung et al., 1992) present an active learning algorithm known as query by committee. In this algorithm, two classifiers are derived from the labeled data and used to label new examples. The instances where the two classifiers disagree are returned to a human annotator for labeling. Lewis and Gale (Lewis and Gale, 1994) pioneered the use of active learning for text categorization. Their scenario, known as pool-based active learning, corresponds to a setting where an abundant supply of text documents is available but only a small sample can be economically annotated by a human labeler. Pool-based active learning has since been explored for many problem domains such as text classification (McCallum and Nigam, 1998; Tong and Chang, 2001; Tong and Koller, 2002), word-sense disambiguation (Chen et al., 2006; Zhu and Hovy, 2007; Dligach and Palmer, 2011), information extraction (Thompson et al., 1999; Settles et al., 2008), and image classification (Tong and Chang, 2001; Hoi et al., 2006).

The pool-based scenario matches the setting in our phenotyping tasks where large supplies of unlabeled EHRs are available but only a small set can be manually reviewed at a reasonable cost. Pool-based active learning is typically an iterative process that operates by first training a classifier on a small sample of the data known as the seed set. The classifier is subsequently applied to a pool of unlabeled data with the purpose of selecting additional examples the classifier views as informative. The selected data is annotated and the cycle is repeated, allowing the learner to quickly refine the decision boundary between classes.

Little research exists on the applications of active learning to the clinical domain. Figueroa et al. (Figueroa et al., 2012) evaluate a Support Vector Machine (SVM) based active learning algorithm in the context of several text classification tasks and find that active learning did not always perform better than random sampling. The use of SVMs restricted their evaluation to binary classification only, limiting the applicability of their findings for many clinical NLP tasks. Chen et al. (Chen et al., 2011) investigate the use of active learning for assertion classification and show that active learning outperforms random sampling. Both of the above mentioned studies experiment with datasets that are quite different from ours in that they annotate relatively short snippets of text. Miller et al. (Miller et al., 2012) develop a series of active learning methods that are highly tailored to coreference resolution in clinical texts. Finally, Hahn et al. (Hahn et al., 2012) utilize active learning in practice for a corpus annotation task that involves labeling pathological phenomena in MEDLINE abstracts. Unfortunately they do not compare the performance of their active learning method to a passive learning baseline, so no conclusion about the effectiveness of active learning can be made. To the best of our knowledge, no work has been done on using active learning for phenotyping. In this work, we experiment with multi-class pool-based active learning in the context of four phenotyping tasks.

3 Methods

3.1 Data Representation

In a phenotyping task, the unit of classification is the patient chart. We represent each chart as a set of Unified Medical Language System (UMLS) (Bodenreider and McCray, 2003) concept identifiers (CUIs) which we extract from the patient records using Apache Clinical Text Analysis and Knowledge Extraction System[1] (cTAKES) (Savova et al., 2010). CUIs aim at abstracting our representations from the lexical variability of medical terminology and capturing the clinically relevant terms in a document leaving out the non-essential and potentially noisy lexical items. Each CUI can be either asserted or negated, as determined by the cTAKES negation module.

Although cTAKES is capable of extracting most CUIs that exist in the UMLS, we only include the CUIs that are listed in phenotype-specific dictionaries. The dictionaries are created manually by domain experts and define the terms that are relevant for each phenotype. Thus, we model each patient $\vec{x}$ as a vector of CUIs where each element n indicates the frequency of occurrence of the respective CUI_n in the records for this patient.

3.2 Models

To perform the classification and to estimate the informativeness of an instance during active learning, we need to evaluate the posterior probability $p(c_i|\vec{x})$, where c_i is the class indicating the relevance of the patient $\vec{x}$ for the given phenotype. For that purpose, we utilize a multinomial Naive Bayes model, which is widely used in document classification.

Naive Bayes classifiers possess several useful properties that make them particularly appropriate for active learning: (1) training and classification speed, (2) ability to produce a probability distribution over the target classes, and (3) ability to perform multi-class classification. Because active learning requires many rounds of retraining (potentially as many as the number of training examples), the first property is crucial for using active learning in practice. The second property is desirable for evaluating the level of uncertainty of the learner over the class predictions. Finally, the third property is important since some of our datasets include more than two classes.

We model the posterior probability as follows:

$$p(c_i|\vec{x}) = \frac{1}{Z} p(c_i) \prod_{n=1}^{N} p(CUI_n|c_i)^{x_n} \quad (1)$$

Where $p(c_i)$ is the prior probability of class c_i, N is the number of CUIs in the phenotype-specific dictionary, CUI_n is the nth CUI in that dictionary, x_n is the frequency of CUI_n in $\vec{x}$, and Z (evidence) is the scaling factor. We determine the model parameters, $p(c_i)$ and $p(CUI_n|c_i)$, using maximum likelihood estimation with Laplace smoothing from the training data. For classification we predict the label c as:

$$c = \arg\max_i p(c_i|\vec{x}) \quad (2)$$

For active learning, we utilize a framework known as *uncertainty sampling* (Lewis and Gale, 1994; Schein and Ungar, 2007). In this framework, the learner requests a label for the instance it is most uncertain how to label. We evaluate the level of uncertainty using the prediction margin metric (Schein and Ungar, 2007) which is defined as:

$$prediction\ margin = |p(c_1|\vec{x}) - p(c_2|\vec{x})| \quad (3)$$

Where c_1 and c_2 are the two most probable classes for the patient $\vec{x}$ according to the model.

3.3 Datasets

In this work we utilize four datasets all of which were created within the i2b2 initiative (Ananthakrishnan et al., 2012; Ananthakrishnan et al., 2013b; Ananthakrishnan et al., 2013a; Xia et al., 2012). We show various important characteristics of our datasets in Table 1. Domain experts defined the ICD-9 codes relevant for each phenotype. These were then used to create the initial cohort from the 6 million+ patient EHR of the Partners Healthcare System. From that initial cohort, 600 patients were randomly chosen for manual labeling.

[1] `http://incubator.apache.org/ctakes/`

Each patient chart was reviewed by a domain expert and labeled at the patient level for CASE or NON-CASE (2-way labeling) for Ulcerative Colitis and Crohn's Disease; CASE, NON-CASE, or UNKNOWN (3-way labeling) for Type II Diabetes; CASE, NON-CASE, PROBABLE, UNKNOWN, or IRRELEVANT (5-way labeling) for Multiple Sclerosis. In our experiments, we used only the clinical narrative data, not a combination of structured and unstructured data. The predominant class for each phenotype was CASE.

3.4 Experimental Setup

Active learning is typically evaluated by comparing the learning curves for passive and active learning-based data selection methods. We generated the learning curves in the style of N fold cross validation ($N = 10$). Within each fold, we have a held out test set and a pool of unlabeled examples. We begin by randomly selecting the seed set of size S, removing it from the pool, and training a model. To produce a point of the active learning curve, we apply the model to the pool of remaining unlabeled data and select the most informative example using the prediction margin metric defined in Equation 3. We move the selected example to the training set, retrain the model, and evaluate its performance on the held out test set. In parallel, to produce a point of the passive learning curve, we select a single example from the pool randomly. We continue this process in an iterative fashion until the pool is exhausted. We repeat this for each of the ten folds and average the resulting learning curves.

In addition, we conduct a series of experiments for each phenotype in which we vary the size of the seed set S. Our motivation is to explore the sensitivity of active learning to the size of the initial seed set. We only try several relatively small seed set sizes. Larger seed sets may erase the gains that could otherwise be obtained by active learning.

In this work, we do not compare the performance of the models accross different phenotypes. Instead, we focus on comparing the performance of active learning against the passive learning baseline.

In practice, active learning is used for selecting examples for subsequent labeling from the pool of unlabeled data. This scenario is simulated in our experiments – we utilize the gold standard data, but we hide the labels from the model. The label is revealed only after the instance is selected and is ready to be added to the training set. This is a common practice used in most published studies of active learning.

4 Results

For each phenotype, we construct the learning curves for different sizes of the seed set. The results are shown in Figures 1, 2, and 3, which include the learning curves for seed sizes $S = 10$, $S = 30$, and $S = 50$ respectively.

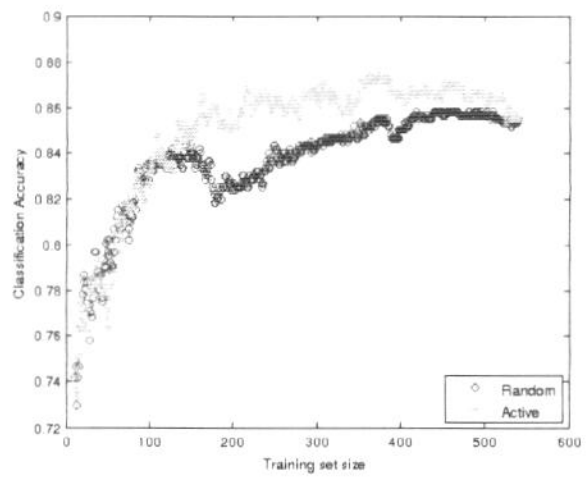

(a) Ulcerative Colitis

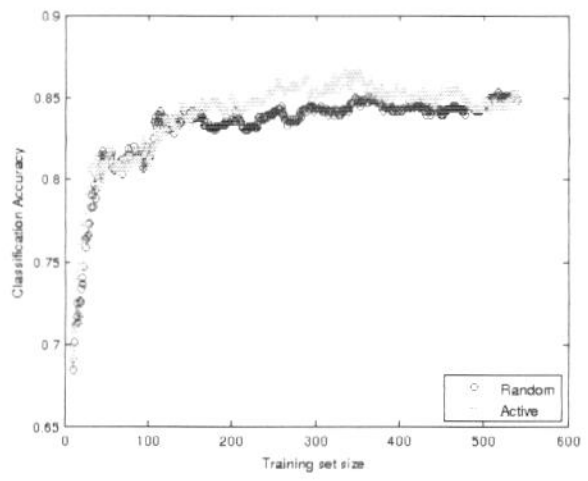

(b) Crohn's Disease

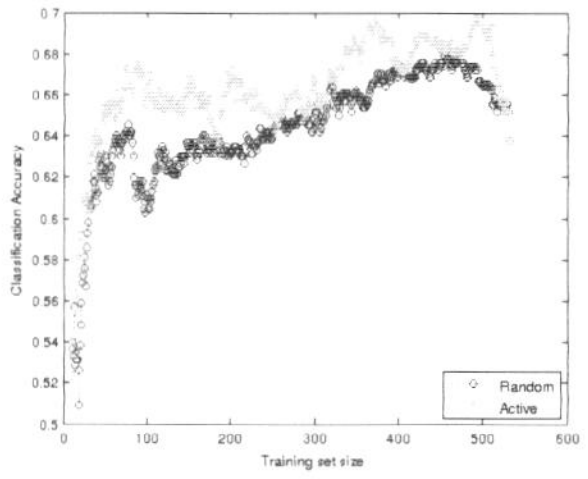

(c) Multiple Sclerosis

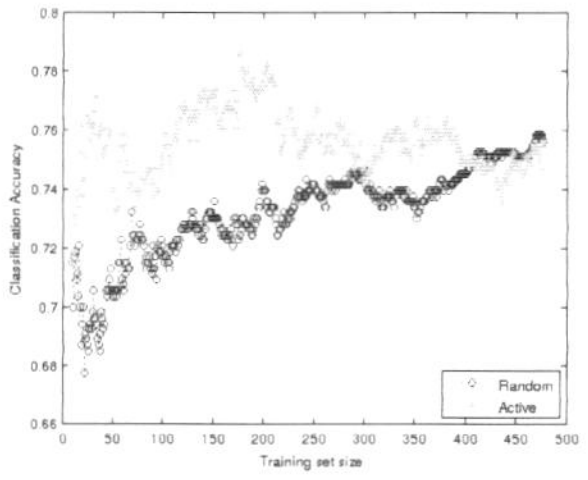

(d) Type II Diabetes

Figure 1: Passive vs. active learning performance on held-out data ($S = 10$)

Phenotype	Total Instances	Number of Classes	Proportion of Predominant Class
Ulcerative Colitis	600	2	0.630
Crohn's Disease	600	2	0.665
Multiple Sclerosis	595	5	0.395
Type II Diabetes	600	3	0.583

Table 1: Dataset Characteristics

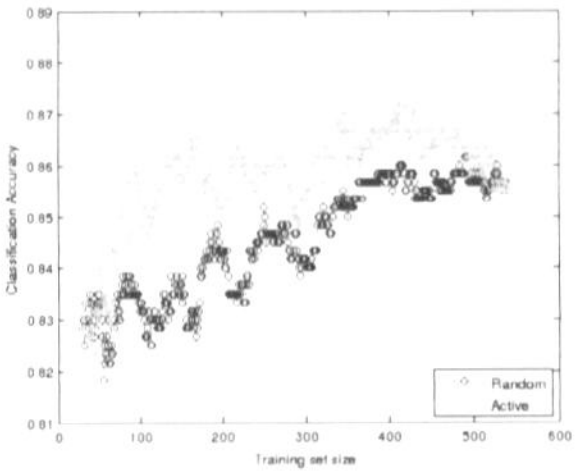

(a) Ulcerative Colitis

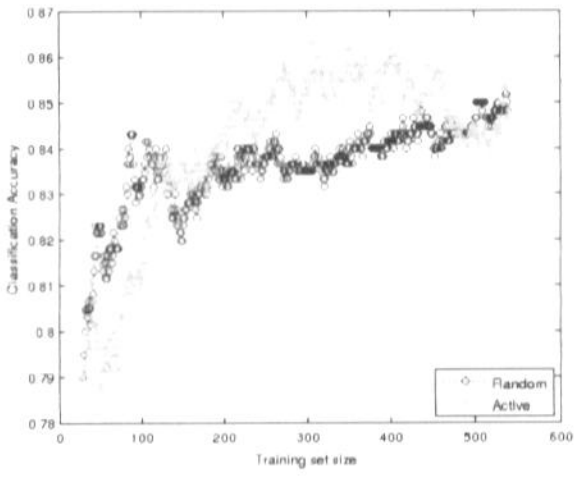

(b) Crohn's Disease

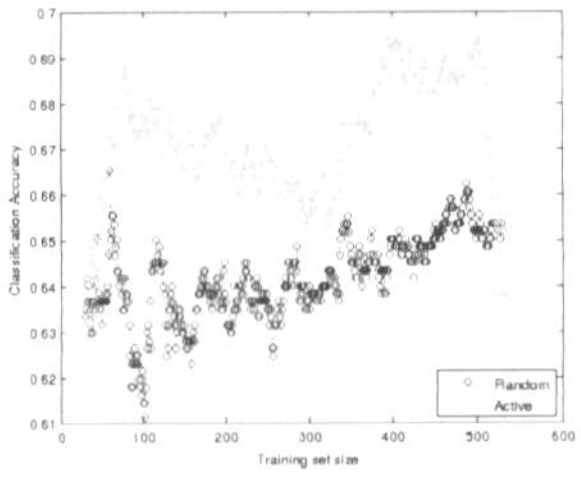

(c) Multiple Sclerosis

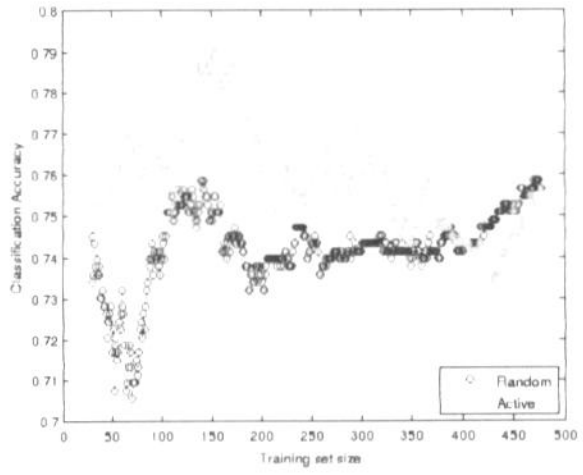

(d) Type II Diabetes

Figure 2: Passive vs. active learning performance on held-out data ($S = 30$)

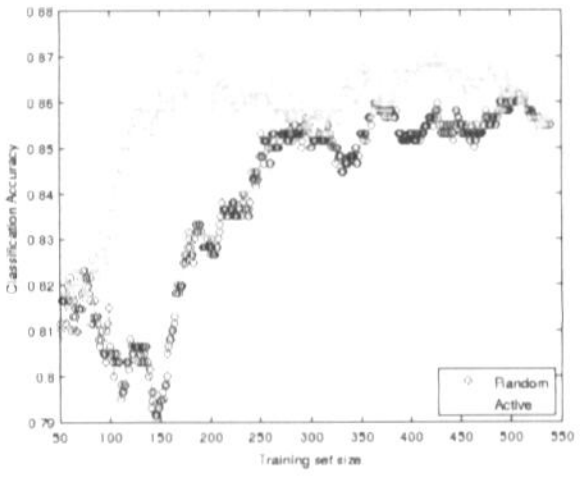

(a) Ulcerative Colitis

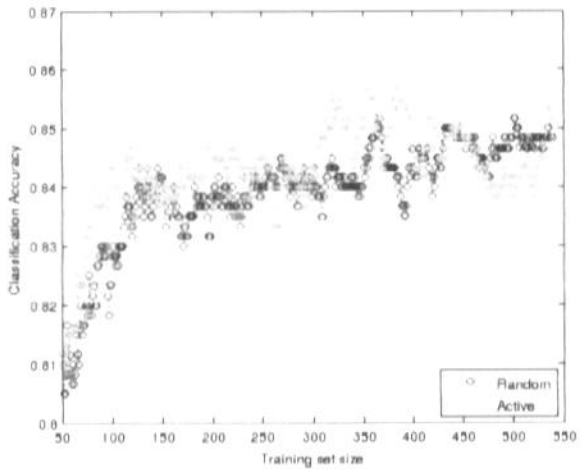

(b) Crohn's Disease

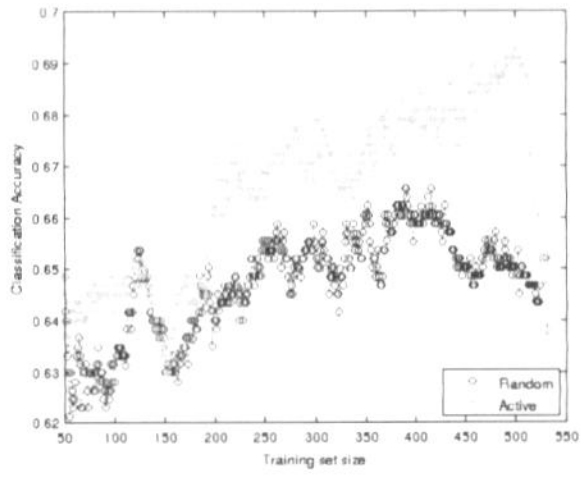

(c) Multiple Sclerosis

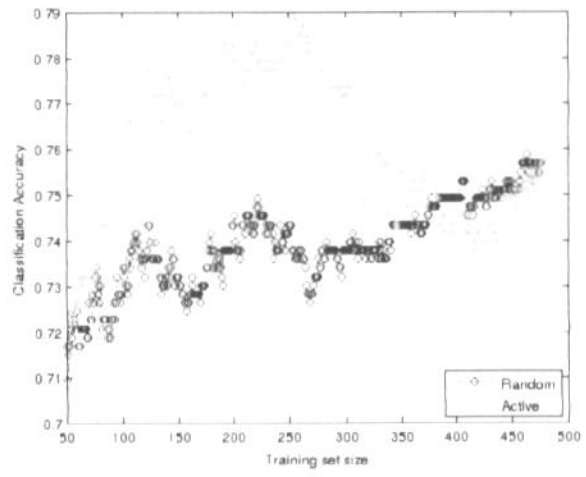

(d) Type II Diabetes

Figure 3: Passive vs. active learning performance on held-out data ($S = 50$)

For each plot we also compute the area under the active learning and passive learning curves. We report the difference between the two curves in Table 2.

5 Discussion and Conclusion

As we see in Figures 1, 2, and 3, for all phenotypes, active learning curves lie above passive

Seed Size	Ulcerative Colitis	Crohn's Disease	Multiple Sclerosis	Type II Diabetes
10	6.90	4.17	10.50	11.05
30	6.64	2.21	15.43	7.49
50	8.63	1.75	8.61	8.90

Table 2: Difference between areas under the curve (Active - Passive)

learning curves for most sizes of the training sets. This means that the models trained on the data selected via active learning typically perform better than the models trained using random sampling. This result is also supported by the fact that the difference between the areas under the active and passive learning curves in Table 2 was positive for all of our experiments.

For most models, active learning reaches the level of the best random sampling performance with fewer than 200 examples, or about 1/3 of the data. This potentially translates into manual annotation savings of about 2/3. Moreover, the best active learning performance is often *above* that of the random sampling baseline. For example, consider Figure 3d. The sizes of the training set in the range between approximately 150 and 350 produce better performance than the best performance of the model trained on the randomly sampled data. At about 260 training examples, the performance of active learning approaches 0.79, which is at least 3 percentage points higher than the performance of the passive learning baseline that it achieves with the *entire* training set.

Although active learning consistently outperformed the passive learning baseline in most of our experiments, occasionally active learning performed worse at certain training set sizes. Consider Figure 2b. During early stages of learning (training set sizes of about 50-130), the passive curve lies above the active learning curve (although active learning recovers later on). We hypothesize that the reason for this behavior lies in outlier selection. Because outliers often do not fit into one of the predefined classes, the classifier is often uncertain about their labels, recommending their selection during active learning. At the same time, the outliers do not help to clarify the decision boundary, negatively affecting the performance. We leave a further investigation into the nature of this behavior for future work.

In other cases, the active learning briefly dips below the passive learning curve at the very end of the selection process. Although this behavior is observed in several cases (e.g. 1d, 2b, 3d), it is unlikely to be of consequence in practice. Active learning would typically be stopped at a much earlier stage, e.g. when 1/3 or 1/2 of the data has been annotated. Nevertheless, it would still be interesting to uncover the conditions leading to this behavior and we leave this investigation for future work.

Both of these scenarios, where active learning performed worse than random sampling, highlight the need for developing stopping criteria for active learning such as (Laws and Schätze, 2008; Bloodgood and Vijay-Shanker, 2009). In a practical application of active learning, a held-out test set is unlikely to be available and some automated means of tracking the progress of active learning is needed. We plan to pursue this avenue of research in the future. In addition to that, we plan to explore the portability of the models trained via active learning. It would also be interesting to investigate the effect of swapping the base classifier: in this work we collect the data for annotation using a multinomial Naive Bayes model. It is still not clear whether the gains obtained by active learning would be preserved if a model was trained on the selected data using a different classifier (e.g. SVM).

Finally, in addition to investigating the performance of active learning across different phenotypes, we also looked at the effects of varying the size of the seed set S. We did not find a clear correlation between the size of the seed set and active learning performance. However, the relationship may exist and could potentially be uncovered if a larger set of seed set sizes was used. We leave the further investigation in this area for future work.

In this work, we explored the use of active learning for several phenotyping tasks. Supervised learning is frequently used for phenotype creation, but the manual annotation that is required for model training is expensive. Active learning offers a way to reduce the annotation costs by involving the classifier in the data selection process. During active learning, the classifier chooses the unlabeled examples it views as informative, thus eliminating the need to annotate the examples that do not contribute to determining the decision boundary. We demonstrated that active learning outperforms the traditionally used passive learning baseline, potentially producing annotation cost savings of up to two-thirds of what is required by the passive baseline.

Acknowledgements

The work described was supported by Temporal History of Your Medical Events (THYME) NLM R01LM010090 and Integrating Informatics and Biology to the Bedside (i2b2) NCBO U54LM008748. Thanks to the anonymous reviewers for thorough and insightful comments. Guergana K. Savova is on the Advisory Board of Wired Informatics, LLC which provides services and products for clinical NLP applications.

References

A.N. Ananthakrishnan, T. Cai, S. Cheng, P.J. Chen, G. Savova, R.G. Perez, V.S. Gainer, S.N. Murphy, P. Szolovits, K. Liao, et al. 2012. Improving case definition of crohn's disease and ulcerative colitis in electronic medical records using natural language processing: a novel informatics approach. *Gastroenterology*, 142(5):S–791.

A. Ananthakrishnan, V. Gainer, T. Cai, Guzman P.R., S. Cheng, G. Savova, P. Chen, P. Szolovits, Z. Xia, P. De Jager, S. Shaw, S. Churchill, E. Karlson, and I. Kohane. 2013a. Similar risk of depression and anxiety following surgery or hospitalization for crohn's disease and ulcerative colitis. *Am J Gastroenterol*.

A.N. Ananthakrishnan, V.S. Gainer, R.G. Perez, T. Cai, S.C. Cheng, G. Savova, P. Chen, P. Szolovits, Z. Xia, P.L. Jager, et al. 2013b. Psychiatric co-morbidity is associated with increased risk of surgery in crohn's disease. *Alimentary pharmacology & therapeutics*, 37(4):445–454.

M. Bloodgood and K. Vijay-Shanker. 2009. A method for stopping active learning based on stabilizing predictions and the need for user-adjustable stopping. In *Proceedings of the Thirteenth Conference on Computational Natural Language Learning*, pages 39–47. Association for Computational Linguistics.

O. Bodenreider and A.T. McCray. 2003. Exploring semantic groups through visual approaches. *Journal of biomedical informatics*, 36(6):414.

J. Chen, A. Schein, L. Ungar, and M. Palmer. 2006. An empirical study of the behavior of active learning for word sense disambiguation. In *Proceedings of the main conference on Human Language Technology Conference of the North American Chapter of the Association of Computational Linguistics*, pages 120–127, Morristown, NJ, USA. Association for Computational Linguistics.

Y. Chen, S. Mani, and H. Xu. 2011. Applying active learning to assertion classification of concepts in clinical text. *Journal of Biomedical Informatics*.

J.C. Denny, M.D. Ritchie, M.A. Basford, J.M. Pulley, L. Bastarache, K. Brown-Gentry, D. Wang, D.R. Masys, D.M. Roden, and D.C. Crawford. 2010. Phewas: demonstrating the feasibility of a phenome-wide scan to discover gene–disease associations. *Bioinformatics*, 26(9):1205–1210.

D. Dligach and M. Palmer. 2011. Good seed makes a good crop: accelerating active learning using language modeling. In *Proceedings of the 49th Annual Meeting of the Association for Computational Linguistics: Human Language Technologies*, volume 2. Association for Computational Linguistics.

R.L. Figueroa, Q. Zeng-Treitler, L.H. Ngo, S. Goryachev, and E.P. Wiechmann. 2012. Active learning for clinical text classification: is it better than random sampling? *Journal of the American Medical Informatics Association*, 19(5):809–816.

U. Hahn, E. Beisswanger, E. Buyko, and E. Faessler. 2012. Active learning-based corpus annotationâîthe pathojen experience. In *AMIA Annual Symposium Proceedings*, volume 2012, page 301. American Medical Informatics Association.

S.C.H. Hoi, R. Jin, J. Zhu, and M.R. Lyu. 2006. Batch mode active learning and its application to medical image classification. In *Proceedings of the 23rd international conference on Machine learning*, pages 417–424. ACM.

A.N. Kho, J.A. Pacheco, P.L. Peissig, L. Rasmussen, K.M. Newton, N. Weston, P.K. Crane, J. Pathak, C.G. Chute, S.J. Bielinski, et al. 2011. Electronic medical records for genetic research: results of the emerge consortium. *Sci Transl Med*, 3(79):79rel.

A.N. Kho, M.G. Hayes, L. Rasmussen-Torvik, J.A Pacheco, W.K. Thompson, L.L. Armstrong, J.C. Denny, P.L Peissig, A.W. Miller, W. Wei, et al. 2012. Use of diverse electronic medical record systems to identify genetic risk for type 2 diabetes within a genome-wide association study. *Journal of the American Medical Informatics Association*, 19(2):212–218.

I.S. Kohane, S.E. Churchill, and S.N. Murphy. 2012. A translational engine at the national scale: informatics for integrating biology and the bedside. *Journal of the American Medical Informatics Association*, 19(2):181–185.

I.J. Kullo, J. Fan, J. Pathak, G.K. Savova, Z. Ali, and C.G. Chute. 2010. Leveraging informatics for genetic studies: use of the electronic medical record to enable a genome-wide association study of peripheral arterial disease. *Journal of the American Medical Informatics Association*, 17(5):568–574.

F. Laws and H. Schätze. 2008. Stopping criteria for active learning of named entity recognition. In *Proceedings of the 22nd International Conference on Computational Linguistics*, volume 1, pages 465–472. Association for Computational Linguistics.

D.D. Lewis and W.A. Gale. 1994. A sequential algorithm for training text classifiers. In *SIGIR '94: Proceedings of the 17th annual international ACM SIGIR conference on Research and development in information retrieval*, pages 3–12, New York, NY, USA. Springer-Verlag New York, Inc.

C. Lin, H. Canhao, T. Miller, D. Dligach, R.M. Plenge, E.W. Karlson, and G. Savova. 2012a. Maximal information coefficient for feature selection for clinical document classification. In *ICML Workshop on Machine Learningfor Clinical Data.*

C. Lin, H. Canhao, T. Miller, D. Dligach, R.M. Plenge, E.W. Karlson, and G.K. Savova. 2012b. Feature engineering and selection for rheumatoid arthritis disease activity classification using electronic medical records. In *ICML Workshop on Machine Learning for Clinical Data Analysis.*

R.M. Long and J.M. Berg. 2011. What to expect from the pharmacogenomics research network. *Clinical Pharmacology & Therapeutics*, 89(3):339–341.

A. McCallum and K. Nigam. 1998. Employing em and pool-based active learning for text classification. In *ICML '98: Proceedings of the Fifteenth International Conference on Machine Learning*, pages 350–358, San Francisco, CA, USA. Morgan Kaufmann Publishers Inc.

C.A. McCarty, R.L. Chisholm, C.G. Chute, I.J. Kullo, G.P. Jarvik, E.B. Larson, R. Li, D.R. Masys, M.D. Ritchie, D.M. Roden, et al. 2011. The emerge network: A consortium of biorepositories linked to electronic medical records data for conducting genomic studies. *BMC medical genomics*, 4(1):13.

T. Miller, D. Dligach, and G Savova. 2012. Active learning for coreference resolution. In *BioNLP: Proceedings of the 2012 Workshop on Biomedical Natural Language Processing*, pages 73–81, Montréal, Canada, June. Association for Computational Linguistics.

F. Olsson. 2009. A literature survey of active machine learning in the context of natural language processing. In *Technical Report, Swedish Institute of Computer Science.*

J.A. Pacheco, P.C. Avila, J.A. Thompson, M. Law, J.A. Quraishi, Alyssa K. Greiman, E.M. Just, and A. Kho. 2009. A highly specific algorithm for identifying asthma cases and controls for genome-wide association studies. In *AMIA Annual Symposium Proceedings*, volume 2009, page 497. American Medical Informatics Association.

G.K. Savova, J.J. Masanz, P.V. Ogren, J. Zheng, S. Sohn, K.C. Kipper-Schuler, and C.G. Chute. 2010. Mayo clinical text analysis and knowledge extraction system (ctakes): architecture, component evaluation and applications. *Journal of the American Medical Informatics Association*, 17(5):507–513.

A.I. Schein and L.H. Ungar. 2007. Active learning for logistic regression: an evaluation. *Machine Learning*, 68(3):235–265.

B. Settles, M. Craven, and S. Ray. 2008. Multiple-instance active learning. *Advances in Neural Information Processing Systems (NIPS)*, 20:1289–1296.

B. Settles. 2009. Active learning literature survey. In *Computer Sciences Technical Report 1648 University of Wisconsin-Madison.*

H. S. Seung, M. Opper, and H. Sompolinsky. 1992. Query by committee. In *COLT '92: Proceedings of the fifth annual workshop on Computational learning theory*, pages 287–294, New York, NY, USA. ACM.

C.A. Thompson, M.E. Califf, and R.J. Mooney. 1999. Active learning for natural language parsing and information extraction. In *Proceedings of the Sixteenth International Conference on Machine Learnin*, pages 406–414. Citeseer.

S. Tong and E. Chang. 2001. Support vector machine active learning for image retrieval. In *Proceedings of the ninth ACM international conference on Multimedia*, pages 107–118. ACM New York, NY, USA.

S. Tong and D. Koller. 2002. Support vector machine active learning with applications to text classification. *The Journal of Machine Learning Research*, 2:45–66.

C. Waudby, R. Berg, J. Linneman, L. Rasmussen, P. Peissig, L. Chen, and C. McCarty. 2011. Cataract research using electronic health records. *BMC ophthalmology*, 11(1):32.

Z. Xia, R. Bove, T. Cai, S. Cheng, R.N.G. Perez, V.S. Gainer, S.N. Murphy, P. Chen, G.K. Savova, K. Liao, E.W. Karlson, S. Shaw, S. Ananthakrishnan, P. Szolovits, S. Churchill, I.S. Kohane, R.M. Plenge, and Philip L.D. 2012. Leveraging electronic health records for research in multiple sclerosis. In *European Committee for Treatment and Research in Multiple Sclerosis (ECTRIMS).*

J. Zhu and E. Hovy. 2007. Active learning for word sense disambiguation with methods for addressing the class imbalance problem. In *Proceedings of the 2007 Joint Conference on Empirical Methods in Natural Language Processing and Computational Natural Language Learning*, pages 783–790.

Finding Negative Symptoms of Schizophrenia in Patient Records

Genevieve Gorrell
The University of Sheffield
g.gorrell@sheffield.ac.uk

Angus Roberts
The University of Sheffield
a.roberts@dcs.shef.ac.uk

Richard Jackson
King's College London
Richard.G.Jackson@slam.nhs.uk

Robert Stewart
King's College London
robert.stewart@kcl.ac.uk

Abstract

This paper reports the automatic extraction of eleven negative symptoms of schizophrenia from patient medical records. The task offers a range of difficulties depending on the consistency and complexity with which mental health professionals describe each. In order to reduce the cost of system development, rapid prototypes are built with minimal adaptation and configuration of existing software, and additional training data is obtained by annotating automatically extracted symptoms for which the system has low confidence. The system was further improved by the addition of a manually engineered rule based approach. Rule-based and machine learning approaches are combined in various ways to achieve the optimal result for each symptom. Precisions in the range of 0.8 to 0.99 have been obtained.

1 Introduction

There is a large literature on information extraction (IE) from the unstructured text of medical records (see (Meystre et al., 2008) for the most recent review). Relatively little of this literature, however, is specific to psychiatric records (see (Sohn et al., 2011; Lloyd et al., 2009; Roque et al., 2011) for exceptions to this). The research presented here helps to fill this gap, reporting the extraction of schizophrenia symptomatology from free text in the case register of a large mental health unit, the South London and Maudsley NHS Trust (SLaM).

We report the extraction of negative symptoms of schizophrenia, such as poor motivation, social withdrawal and apathy. These often present in addition to more prominent, positive symptoms such as delusions and hallucinations. Negative symptoms can severely impair the quality of life of affected patients, yet existing antipsychotic medications have poor efficacy in their treatment. As negative symptoms can be measured in quantitative frameworks within a clinical environment (Kay et al., 1987; Andreasen, 1983), they have the potential to reflect the success or failure of new medical interventions, and are of widespread interest in the epidemiology of schizophrenia. The motivation for our work is to provide information on the presence of negative symptoms, for use in such quantitative measures.

SLaM covers a population of 1.1 million, being responsible for close to 100% of the mental health care contacts in four London boroughs. Approximately 225,000 records are stored in the SLaM Electronic Health Record (EHR) system, which supports an average of 35,000 patients at any one time. SLaM hosts the UK National Institute for Health Research (NIHR) Biomedical Research Center (BRC) for Mental Health. The BRC deidentifies all records in the SLaM EHR (Fernandes et al., 2013) to form the largest mental health case register in Europe, the Case Register Interactive Search (CRIS) system (Stewart et al., 2009). CRIS provides BRC epidemiologists with search facilities, via a web front end that allows standard information retrieval queries over an inverted index, and via database query languages. CRIS has been approved as an anonymized data resource for secondary analysis by Oxfordshire Research Ethics Committee C (08/H0606/71). The governance for all CRIS projects and dissemination is managed through a patient-led oversight committee.

CRIS contains both the structured information, and the unstructured free text from the SLaM

EHR. The free text consists of 18 million text field instances – a mix of correspondence and notes describing patient encounters. Much of the information of value to mental health epidemiologists is found in these free text fields. SLaM clinicians record important information in the textual portion of the record, even when facilities are provided for recording the same information in a structured format. For example, a query on the structured fields containing Mini Mental State Examination scores (MMSE, a score of cognitive ability) recently returned 5,700 instances, whereas a keyword search over the free text fields returned an additional 48,750 instances. The CRIS inverted index search system, however, cannot return the specific information of interest (the MMSE score in this case), instead returning each text field that contains a query match, in its entirety. In the case of symptomatology, as examined in this paper, symptoms are rarely recorded in structured fields, but are frequently mentioned in the unstructured text.

This problem is not unusual. (Meystre et al., 2008) note that free text is "convenient to express concepts and events" (Meystre et al., 2008), but that it is difficult for re-use in other applications, and difficult for statistical analysis. (Rosenbloom et al., 2011) have reviewed the few studies that look at the expressivity of structured clinical documentation systems compared to natural prose notes, and report that prose is more accurate, reliable and understandable. (Powsner et al., 1998) refer to structured data as freezing clinical language, and restricting what may be said. (Greenhalgh et al., 2009), referring to the free text of the paper record, say that it is tolerant of ambiguity, which supports the complexity of clinical practice. Much of medical language is hedged with ambiguity and probability, which is difficult to represent as structured data (Scott et al., 2012).

Given the presence of large quantities of valuable information in the unstructured portion of the BRC case register, and CRIS's inability to extract this information using standard information retrieval techniques, it was decided, in 2009, to implement an IE and text mining capability as a component of CRIS. This comprises tools to develop and evaluate IE applications for specific end-user requirements as they emerge, and the facility to deploy these applications on the BRC compute cluster.

Most IE applications developed by the BRC to date have used a pattern matching approach. In this, simple lexico-syntactic pre-processing and dictionary lookup of technical terms are followed by cascades of pattern matching grammars designed to find the target of extraction. These grammars are hand-written by language engineers. Previous extraction targets have included smoking status, medications, diagnosis, MMSE, level of education, and receipt of social care. Building such pattern matching grammars is often time consuming, in that it takes significant language engineer time to develop and refine grammars. In addition, the process of writing and testing grammars requires examples of the extraction target. These are provided by manual annotation, or labelling, of examples and correction of system output; a task which takes significant domain expert time.

In the case of schizophrenia, the IE applications are required to extract multiple symptoms for use in quantitative measures of the disease. The set of symptoms relevant to such quantitative measures number in the dozens. Given the cost of pattern grammar development, and the cost of manual annotation, it is impractical to develop grammars for each of the required symptoms, and such an approach would not scale up to larger numbers of symptoms and to other diseases. In addition, the cost of domain expert annotation of examples for each individual symptom is also high. The approach taken in our research aims to reduce these two costs.

In order to reduce the cost of system development, and to improve scalability to new symptoms and diseases, we build rapid prototypes, using off-the-shelf NLP and machine learning (ML) toolkits. Such toolkits, and repositories of applications built on them, are becoming increasingly popular. It has been asked (Nadkarni et al., 2011) whether such tools may be used as "commodity software" to create clinical IE applications with little or no specialist skills. In order to help answer this question, we compare the performance of our ML only prototypes to applications that combine ML and pattern matching, and to applications implemented with pattern matching alone.

The second cost considered is that of finding and labelling high quality examples of the extraction target, used to inform and test system development. To deal with this cost, we explore methods of enriching the pool of examples for labelling, including the use of methods inspired by active learning (Settles, 2012). In active learning, potential examples of the extraction target are selected by the learning algorithm for labelling by the human annotator. The aim is to present instances which will most benefit the ML algorithm, at least human cost. This paper presents results from experiments in training data enrichment, and a simple approach to active learning, applied to symptom extraction.

The paper is organised as follows. Section 2 looks at the task domain in more detail, explaining the symptoms to be extracted, and describing the dataset. Section 3 describes the experimental method used, and the evaluation metrics. This is followed by a presentation of the results in Section 4, and a discussion of these results in Section 5. Finally, we draw some conclusions in Section 6.

2 Analysis of the Task Domain

In this section we will first introduce the concept of negative symptoms and explain what entities we are aiming to extract from the data. We will then discuss the datasets we used, and how each symptom varies in its nature and therefore difficulty.

2.1 Negative Symptoms

In the psychiatric context, negative symptoms are deficit symptoms; those that describe an absence of a behaviour or ability that would normally be present. A positive symptom would be one which is not normally present. In schizophrenia, positive symptoms might include delusions, auditory hallucinations and thought disorder. Here, we are concerned with negative symptoms of schizophrenia, in particular the following eleven, where bold font indicates the feature values we hope to extract from the data (in machine learning terms, the classes, not including the negative class). Examples illustrate something of the ways in which the symptom might be described in text. "ZZZZZ" replaces the patient name for anonymization purposes:

- **Abstract Thinking**: Does the individual show evidence of requiring particularly **concrete** conceptualizations in order to understand? Examples include; "Staff have noted ZZZZZ is very concrete in his thinking", "Thought disordered with concrete thinking", but NOT "However ZZZZZ has no concrete plans to self-harm"

- **Affect**: Is the individual's emotional response **blunted** or **flat**? Is it inappropriate to events (**abnormal**)? Alternatively, does the individual respond appropriately (**reactive**)? Examples include; "Mood: subjectively 'okay' however objectively incongruent", "Denied low mood or suicide ideation", "showed blunting of affect"

- **Apathy**: Does the individual exhibit **apathy**? Examples include; "somewhat apathetic during his engagement in tasks", "Apathy."

- **Emotional Withdrawal**: Does the individual appear **withdrawn** or **detached**? Examples include; "withdrawal from affectational and social contacts", "has been a bit withdrawn recently", NOT "socially withdrawn", which is a separate symptom, described below.

- **Eye Contact**: Does the individual make **good** eye contact, or is it **intermediate** or **poor**? Examples include; "eye contact was poor", "maintaining eye contact longer than required", "made good eye contact"

- **Motivation**: Is motivation **poor**? Examples include; "ZZZZZ struggles to become motivated.", "ZZZZZ lacks motivation.", "This is due to low motivation."

- **Mutism**: A more extreme version of poverty of speech (below), and considered a separate symptom, is the individual **mute** (but not deaf mute)? Examples include; "Was electively mute [...]", "ZZZZZ kept to himself and was mute.", NOT "ZZZZZ is deaf mute."

- **Negative Symptoms**: An umbrella term for the symptoms described here. Do we see any **negative symptom**? Examples include; "main problem seems to be negative symptoms [...]", "[...] having negative symptoms of schizophrenia."

- **Poverty of Speech**: The individual may show a deficit or **poverty** of speech, or their speech may be **abnormal** or **normal**. Examples include; "Speech: normal rate and rhythm", "speech aspontaneous", "speech was dysarthric", "ongoing marked speech defect", "speech was coherent and not pressured"

- **Rapport**: Individual ability to form conversational rapport may be **poor** or **good**. Examples include; "we could establish a good rapport", "has built a good rapport with her carer"

- **Social Withdrawal**: Do we see indications of **social withdrawal** or not? Examples include; "long term evidence of social withdrawal", "ZZZZZ is quite socially withdrawn"

2.2 Dataset

Different symptoms vary in the challenges they pose. For example, "apathy" is almost exclusively referred to using the word "apathy" or "apathetic", and where this word appears, it is almost certainly a reference to the negative symptom of apathy, whereas concrete thinking is harder to locate because the word "concrete" appears so often in other contexts, and because concrete thinking may be referred to in less obvious ways. In the previous section, we gave some examples of negative symptom mentions that give an idea of the range of possibilities. Exemplars were unevenly distributed among medical records, with some records having several and others having none.

Due to the expertize level required for the annotation part of the task, and strict limitations on who is authorized to view the data, annotation was performed by a single psychiatrist. Data quantity was therefore limited by the amount of time the expert annotator had available for the work. For this reason, formal interannotator agreement assessment was not possible, although a second annotator did perform some consistency checking on the data. Maximizing the utility of a limited dataset therefore constituted an important part of the work.

Because many of the records do not contain any mention of the symptom in question, in order to make a perfect gold standard corpus the expert annotator would have to read a large number of potentially very lengthy documents looking for mentions that are thin on the ground. Because expert annotator time was so scarce, this was likely to lead to a much reduced corpus size, and so a compromise was arrived at whereby simple heuristics were used to select candidate mentions for the annotator to judge rather than having also to find them. For example, in abstract thinking, one heuristic used was to identify all mentions of "concrete". In some cases, the mention is irrelevant to concrete thinking, so the annotator marks it as a negative, whereas in others it is a positive mention. This means that compared with a fully annotated corpus, our data may be lower on recall, since some cases may not have been identified using the simple heuristics, though precision is most likely excellent, since all positive examples have been fully annotated by the expert. In terms of the results reported here, this compromise has little impact, since the task is defined to be replicating the expert annotations, whatever they may be. However, it might be suggested that our task is a little easier than it would have been for a fully annotated corpus, since the simple heuristics used to identify mentions would bias the task toward the easier cases. In terms of the adequacy of the result for future use cases, precision is the priority so this decision was made with end use in mind.

2.2.1 Selecting examples for training

As a further attempt to obtain more expert-annotated data, the principles of active learning were applied in order to strategically leverage annotator time on the most difficult cases and for the most difficult symptoms. Candidate mentions were extracted with full sentence context on the basis of their confidence scores, as supplied by the classifier algorithm, and presented to the annotator for judgement. Mentions were presented in reverse confidence score order, so that annotator time was prioritized on those examples where the classifier was most confused.

3 Method

Because the boundaries of a mention of a negative symptom are somewhat open to debate, due to the wide variety of ways in which psychiatric professionals may describe a negative symptom, we defined the boundaries to be sentence boundaries, thus transforming it into a sentence classification task. However, for evaluation purposes, precision, recall and F1 are used here, since observed agreement is not appropriate for an entity extraction task, giving an inflated result due to the inevitably large number of correctly classified negative examples.

Due to the requirements of the use case, our work was biased toward achieving a good precision. Future work making use of the data depends upon the results being of good quality, whereas a lower recall will only mean that a smaller proportion of the very large amount of data is available. For this reason, we aimed, where possible, to achieve precisions in the region of 0.9 or higher, even at the expense of recalls below 0.6.

Our approach was to produce a rapid prototype with a machine learning approach, and then to combine this with rule-based approaches in an attempt to improve performance. Various methods of combining the two approaches were tried. Machine learning alone was performed using support vector machines (SVMs). Two rule phases were then added, each with a separate emphasis on improving either precision or recall. The rule-based approach was then tried in the absence of a machine learning component, and in addition both overriding the ML where it disagreed and being overridden by it. Rules were created using the JAPE language (Cunningham et al., 2000). Experiments were performed using GATE (Cunningham et al., 2011; Cunningham et al., 2013), and the SVM implementation provided with GATE (Li et al., 2009).

Evaluation was performed using fivefold cross-validation, to give values for precision, recall and F1 using standard definitions. For some symptoms, active learning data were available (see Section 2.2.1) comprising a list of examples chosen for having a low confidence score on earlier versions of the system. For these symptoms, we first give a result for systems trained on the original dataset. Then, in order to evaluate the impact of this intervention, we give results for systems trained on data including the specially selected data. However, at test time, these data constitute a glut of misrepresentatively difficult examples that would have given a deflated result. We want to include these only at training time and not at test time. Therefore, the fold that contained these data in the test set was excluded from the calculation. For these symptoms, evaluation was based on the four out of five folds where the active learning data fell in the training set. The symptoms to which this applies are abstract thinking, affect, emotional withdrawal, poverty of speech and rapport.

In the next section, results are presented for these experiments. The discussion section focuses on how results varied for different symptoms, both in the approach found optimal and the result achieved, and why this might have been the case.

4 Results

Table 1 shows results for each symptom obtained using an initial "rapid prototype" support vector machine learner. Confidence threshold in all cases is 0.4 except for negative symptoms, where the confidence threshold is 0.6 to improve precision. Features used were word unigrams in the sentence in conjunction with part of speech (to distinguish for example "affect" as a noun from "affect" as a verb) as well as some key terms flagged as relevant to the domain. Longer n-grams were rejected as a feature due to the small corpus sizes and consequent risk of overfitting. A linear kernel was used. The soft margins parameter was set to 0.7, allowing some strategic misclassification in boundary selection. An uneven margins parameter was used (Li and Shawe-Taylor, 2003; Li et al., 2005) and set to 0.4, indicating that the boundary should be positioned closer to the negative data to compensate for uneven class sizes and guard against small classes being penalized for their rarity. Since the amount of data available was small, we were not able to reserve a validation set, so care was taken to select parameter values on the basis of theory rather than experimentation on the test set, although confidence thresholds were set pragmatically. Table 1 also gives the number of classes, including the negative class (recall that different symptoms have different numbers of classes), and number of training examples, which give some information about task difficulty.

As described in Section 2.2.1, active learning-style training examples were also included for symptoms where it was deemed likely to be of benefit. Table 2 provides performance statistics for these symptoms alongside the original machine learning result for comparison. In all cases, some improvement was observed, though the extent of the improvement was highly variable.

Central to our work is investigating the interplay between rule-based and machine learning approaches. Rules were prepared for most symptoms, with the intention that they should be complementary to the machine learning system, rather than a competitor. The emphasis with the rules is on coding for the common patterns in both positive and negative examples, though coding the ways in which a symptom might not be referred

Table 1: Machine Learning Only, SVM

Symptom	Classes	Training Ex.	Precision	Recall	F1
Abstract Thinking	2	118	0.615	0.899	0.731
Affect	5	103	0.949	0.691	0.8
Apathy	2	145	0.880	0.965	0.921
Emotional Withdrawal	3	118	0.688	0.815	0.746
Eye Contact	4	35	0.827	0.677	0.745
Motivation	2	259	0.878	0.531	0.662
Mutism	2	234	0.978	0.936	0.956
Negative Symptoms	2	185	0.818	0.897	0.856
Poverty of Speech	4	263	0.772	0.597	0.674
Rapport	3	139	0.775	0.693	0.731
Social Withdrawal	2	166	0.940	0.958	0.949

Table 2: Active Learning

Symptom	Ex.	Without AL-Style Examples			With AL-Style Examples			Difference
		Prec	Rec	F1	Prec	Rec	F1	
Abstract Thinking	99	0.595	0.940	0.728	0.615	0.899	0.731	0.003
Affect	200	0.947	0.529	0.679	0.949	0.691	0.8	0.121
Emotional Withdrawal	100	0.726	0.517	0.604	0.688	0.815	0.746	0.142
Poverty of Speech	62	0.721	0.515	0.601	0.772	0.597	0.674	0.073
Rapport	37	0.725	0.621	0.669	0.775	0.693	0.731	0.062

to is considerably harder. F1 results for the standalone rule-based systems where sufficiently complete are given in Table 4; however, for now, we focus on the results of our experiments in combining the two approaches, which are given in Table 3. Here, we give results for layering rules with machine learning. On the left, we see results obtained where ML first classifies the examples, then the rule-based approach overrides any ML classification it disagrees with. In this way, the rules take priority. On the right, we see results obtained where machine learning overrides any rule-based classification it disagrees with. The higher of the F1 scores is given in bold. Results suggest that the more successful system is obtained by overriding machine learning with rules rather than vice versa.

Table 4 gives a summary of the best results obtained by symptom, using all training data, including active learning instances. We focus on F1 scores only here for conciseness. The baseline machine learning result is first recapped, along with the rule-based F1 where this was sufficiently complete to stand alone. Since in all cases, overriding machine learning with rules led to the best result of the two combination experiments, we give the F1 for this, which in all cases, where available, proves the best result of all. We provide the percentage improvement generated relative to the ML baseline by the combined approach. The final column recaps the best F1 obtained for that symptom. We can clearly see from Table 4 that in all cases, the result obtained from combining approaches outperforms either of the approaches taken alone.

5 Discussion

In summary, the best results were obtained by building upon a basic SVM system with layers of rules that completed and corrected areas of weakness in the machine learning. Note that the symptoms where this approach yielded the most striking improvements tended to be those with the fewer training examples and the larger numbers of classes. In these cases, the machine learning approach is both easier to supplement using rules and easier to beat. A high performing rule-based system certainly correlates with a substantial improvement over the ML baseline; however, we

Table 3: Machine Learning Layered with Rules

Symptom	Rules Override ML			ML Overrides Rules		
	Precision	Recall	F1	Precision	Recall	F1
Abstract Thinking	0.914	0.719	**0.805**	0.935	0.652	0.768
Affect	0.931	0.827	0.876	0.931	0.827	0.876
Emotional Withdrawal	0.840	0.778	**0.808**	0.691	0.827	0.753
Eye Contact	0.88	0.852	**0.866**	0.779	0.611	0.684
Mutism	0.986	0.936	**0.960**	0.978	0.936	0.956
Negative Symptoms	0.851	0.897	**0.874**	0.818	0.897	0.856
Poverty of Speech	0.8	0.730	**0.763**	0.793	0.723	0.757
Rapport	0.839	0.868	**0.853**	0.907	0.772	0.834

Table 4: Best Result Per Symptom

Symptom	Classes	Ex.	ML F1	Rules F1	Rules>ML F1	% Imp	Best F1
Abstract Thinking	2	217	0.731	0.765	0.805	10%	0.805
Affect	5	303	0.800	0.820	0.876	9%	0.876
Apathy	2	145	0.921	n/a	n/a	n/a	0.921
Emotional withdrawal	3	218	0.746	0.452	0.808	8%	0.808
Eye contact	4	35	0.745	0.859	0.866	16%	0.866
Motivation	2	259	0.662	n/a	n/a	n/a	0.662
Mutism	2	234	0.956	n/a	0.960	0%	0.960
Negative Symptoms	2	185	0.856	n/a	0.874	2%	0.874
Poverty of speech	4	325	0.674	0.689	0.763	13%	0.763
Rapport	3	176	0.731	0.826	0.853	17%	0.853
Social withdrawal	2	166	0.949	n/a	n/a	n/a	0.949

do also consistently see the combined approach outperforming both the ML and rule-based approaches as taken separately. We infer that this approach is of the most value in cases where training data is scarce.

Where machine learning was removed completely, we tended to see small performance decreases, but in particular, recall was badly affected. Precision, in some cases, improved, but not by as much as recall decreased. This seems to suggest that where datasets are limited, machine learning is of value in picking up a wider variety of ways of expressing symptoms. Of course, this depends on a) the coverage of the rules against which the SVM is being contrasted, and b) the confidence threshold of the SVM and other relevant parameters. However, this effect persisted even after varying the confidence threshold of the SVM quite substantially.

Optimizing precision presented more difficulties than improving recall. Varying the confidence threshold of the SVM to improve recall tended to cost more in recall than was gained in precision, so rule-based approaches were employed. However, it is much easier to specify what patterns do indicate a particular symptom than list all the ways in which the symptom might *not* be referred to. Symptoms varied a lot with respect to the extent of the precision problem. In particular, abstract thinking, which relies a lot on the word "concrete", which may appear in many contexts, posed problems, as did emotional withdrawal, which is often indicated by quite varied use of the word "withdrawn", which may occur in many contexts. Other symptoms, whilst easier than abstract thinking and social withdrawal, are also variable in the way they are expressed. Mood, for example, is often described in expressive and indirect ways, as is poverty of speech. On the other hand, mutism is usually very simply described, as is eye contact. It is an aid in this task that medical professionals often use quite formalized and predictable ways of referring to symptoms.

Aside from that, task difficulty depended to a large extent on the number of categories into which symptoms may be split. For example, the simple "mute" category is easier than eye contact, which may be good, intermediate or poor, with intermediate often being difficult to separate from good and poor. Likewise, speech may show poverty or be normal or abnormal, with many different types of problem indicating abnormality.

We chose to use an existing open-source language engineering toolkit for the creation of our applications; namely GATE (Cunningham et al., 2011). This approach enabled rapid prototyping, allowing us to make substantial progress on a large number of symptoms in a short space of time. The first version of a new symptom was added using default tool settings and with no additional programming. It was often added to the repertoire in under an hour, and although not giving the best results, this did achieve a fair degree of success, as seen in Table 1 which presents the machine learning-only results. In the case of the simpler symptoms (apathy and social withdrawal), this initial system gave sufficient performance to require no further development.

Additional training data was obtained for five symptoms, by presenting labelled sentences with low classifier confidence to the annotator (Table 2). Although this did improve performance, it is unclear whether this was due to an increase in training data alone, or whether concentrating on the low confidence examples made a difference. The annotator did, however, report that they found this approach easier, and that it took less time than annotating full documents for each symptom.

6 Conclusion

In conclusion, a good degree of success has been achieved in finding and classifying negative symptoms of schizophrenia in medical records, with precisions in the range of 0.8 to 0.99 being achieved whilst retaining recalls in excess of 0.5 and in some cases as high as 0.96. The work has unlocked key variables that were previously inaccessible within the unstructured free text of clinical records. The resulting output will now feed into epidemiological studies by the NIHR Biomedical Research Centre for Mental Health.

We asked whether off-the-shelf language engineering software could be used to build symptom extraction applications, with little or no additional configuration. We found that it is possible to create prototypes using such a tool, and that in the case of straightforward symptoms, these perform well. In the case of other symptoms, however, language engineering skills are required to enhance performance. The best results were obtained by adding hand-crafted rules that dealt with weakness in the machine learning.

References

N. C. Andreasen. 1983. *Scale for the Assessment of Negative Symptoms*. University of Iowa Press, Iowa City. Cited by 0000.

H. Cunningham, D. Maynard, and V. Tablan. 2000. JAPE: a Java Annotation Patterns Engine (Second Edition). Research Memorandum CS–00–10, Department of Computer Science, University of Sheffield, Sheffield, UK, November.

H. Cunningham, D. Maynard, K. Bontcheva, V. Tablan, N. Aswani, I. Roberts, G. Gorrell, A. Funk, A. Roberts, D. Damljanovic, T. Heitz, M.A. Greenwood, H. Saggion, J. Petrak, Y. Li, and W. Peters. 2011. *Text Processing with GATE (Version 6)*.

Hamish Cunningham, Valentin Tablan, Angus Roberts, and Kalina Bontcheva. 2013. Getting more out of biomedical documents with gate's full lifecycle open source text analytics. *PLoS Computational Biology*, 9(2):e1002854, 02.

Andrea C Fernandes, Danielle Cloete, Matthew TM Broadbent, Richard D Hayes, Chin-Kuo Chang, Angus Roberts, Jason Tsang, Murat Soncul, Jennifer Liebscher, Richard G Jackson, Robert Stewart, and Felicity Callard. 2013. Development and evaluation of a de-identification procedure for a case register sourced from mental health electronic records. *BMC Medical Informatics and Decision Making*. Accepted for publication.

T. Greenhalgh, H. W. Potts, G. Wong, P. Bark, and D. Swinglehurst. 2009. Tensions and paradoxes in electronic patient record research: a systematic literature review using the meta-narrative method. *Milbank Quarterly*, 87(4):729–788, Dec.

S R Kay, A Fiszbein, and L A Opler. 1987. The positive and negative syndrome scale (PANSS) for schizophrenia. *Schizophrenia bulletin*, 13(2):261–276. Cited by 8221.

Y. Li and J. Shawe-Taylor. 2003. The SVM with Uneven Margins and Chinese Document Categorization. In *Proceedings of The 17th Pacific Asia Conference on Language, Information and Computation (PACLIC17)*, Singapore, Oct.

Y. Li, K. Bontcheva, and H. Cunningham. 2005. Using Uneven Margins SVM and Perceptron for Information Extraction. In *Proceedings of Ninth Conference on Computational Natural Language Learning (CoNLL-2005)*.

Yaoyong Li, Kalina Bontcheva, and Hamish Cunningham. 2009. Adapting SVM for Data Sparseness and Imbalance: A Case Study on Information Extraction. *Natural Language Engineering*, 15(2):241–271.

Keith Lloyd, Matteo Cella, Michael Tanenblatt, and Anni Coden. 2009. Analysis of clinical uncertainties by health professionals and patients: an example from mental health. *BMC Medical Informatics and Decision Making*, 9(1):34.

S.M. Meystre, G.K. Savova, K.C. Kipper-Schuler, and J.F. Hurdle. 2008. Extracting information from textual documents in the electronic health record: A review of recent research. *IMIA Yearbook of Medical Informatics*, pages 128–144.

P. M. Nadkarni, L. Ohno-Machado, and W. W. Chapman. 2011. Natural language processing: an introduction. *J Am Med Inform Assoc*, 18(5):544–551.

S. M. Powsner, J. C. Wyatt, and P. Wright. 1998. Opportunities for and challenges of computerisation. *Lancet*, 352(9140):1617–1622, Nov.

Francisco S. Roque, Peter B. Jensen, Henriette Schmock, Marlene Dalgaard, Massimo Andreatta, Thomas Hansen, Karen Seby, Sren Bredkjr, Anders Juul, Thomas Werge, Lars J. Jensen, and Sren Brunak. 2011. Using electronic patient records to discover disease correlations and stratify patient cohorts. *PLoS Comput Biol*, 7(8):e1002141, 08.

S. T. Rosenbloom, J. C. Denny, H. Xu, N. Lorenzi, W. W. Stead, and K. B. Johnson. 2011. Data from clinical notes: a perspective on the tension between structure and flexible documentation. *J Am Med Inform Assoc*, 18(2):181–186.

Donia Scott, Rossano Barone, and Rob Koeling. 2012. Corpus annotation as a scientific task. In Nicoletta Calzolari (Conference Chair), Khalid Choukri, Thierry Declerck, Mehmet Uur Doan, Bente Maegaard, Joseph Mariani, Jan Odijk, and Stelios Piperidis, editors, *Proceedings of the Eight International Conference on Language Resources and Evaluation (LREC'12)*, Istanbul, Turkey, may. European Language Resources Association (ELRA).

Burr Settles. 2012. *Active Learning*. Morgan and Claypool.

Sunghwan Sohn, Jean-Pierre A Kocher, Christopher G Chute, and Guergana K Savova. 2011. Drug side effect extraction from clinical narratives of psychiatry and psychology patients. *Journal of the American Medical Informatics Association*, 18(Suppl 1):i144–i149.

Robert Stewart, Mishael Soremekun, Gayan Perera, Matthew Broadbent, Felicity Callard, Mike Denis, Matthew Hotopf, Graham Thornicroft, and Simon Lovestone. 2009. The South London and Maudsley NHS foundation trust biomedical research centre (SLAM BRC) case register: development and descriptive data. *BMC Psychiatry*, 9:51–62.

De-Identification of Clinical Free Text in Dutch with Limited Training Data: A Case Study

Elyne Scheurwegs
Artesis Hogeschool Antwerpen
elynescheurwegs@hotmail.com

Kim Luyckx
biomina - biomedical informatics group
Antwerp University Hospital
& University of Antwerp
kim.luyckx@uza.be

Filip Van der Schueren
Artesis Hogeschool Antwerpen
filip.vanderschueren@artesis.be

Tim Van den Bulcke
biomina - biomedical informatics group
Antwerp University Hospital
& University of Antwerp
tim.vandenbulcke@uza.be

Abstract

In order to analyse the information present in medical records while maintaining patient privacy, there is a basic need for techniques to automatically de-identify the free text information in these records. This paper presents a machine learning de-identification system for clinical free text in Dutch, relying on best practices from the state of the art in de-identification of English-language texts. We combine string and pattern matching features with machine learning algorithms and compare performance of three different experimental setups using Support Vector Machines and Random Forests on a limited data set of one hundred manually obfuscated texts provided by Antwerp University Hospital (UZA). The setup with the best balance in precision and recall during development was tested on an unseen set of raw clinical texts and evaluated manually at the hospital site.

1 Introduction

In Electronic Health Records (EHRs), medical information about the treatment of patients is stored on a daily basis, both in structured (e.g. lab results, medication,) and unstructured (e.g. clinical notes) forms. EHRs are unique sources of information that need be further analyzed to improve diagnosis and treatment of future patients. However, these information sources cannot be freely explored due to privacy regulations (Privacy Rule, 2002; European Data Protection Directive, 1995; Belgian Data Protection Act, 1993). Automated de-identification is crucial to remove personal health information (PHI), while keeping all medical and contextual information as intact as possible. In the US, this is regulated under the Health Insurance Portability and Accountability Act (HIPAA, 1996).

Approaches to de-identification can be categorised into two main types, with rule-based and pattern matching approaches on the one hand and machine learning approaches on the other, as suggested in Meystre et al. (2010). Rule-based and pattern-matching approaches often rely on dictionaries and manually constructed regular expressions. While this type of approach does not require any annotation effort and can easily be customised to increase performance, it offers only limited scalability and is often highly language dependent. Machine learning approaches in general are better scalable and more robust to noise, but especially supervised learning algorithms require substantial amounts of annotated training data, a very time-consuming and expensive undertaking. The selection of meaningful features is a crucial aspect in the machine learning approach, especially when only limited data is available (Ferrández et al., 2012a). Hybrid approaches to de-identification such as that presented in Ferrández et al. (2012b) have been developed to combine the advantages of the machine learning approach with those of dictionaries and regular expressions. Below, we highlight a number of interesting studies from the state of the art in automated de-identification.

One of the first systems for de-identification, the Scrub system, was proposed in Sweeney et al. (1996). Scrub takes a dictionary rule-based approach and has been shown to be able to effectively model the human approach to locating PHI entities. This study included well-formatted letters with a header block as well as shorthand notes, but does not provide details on recall and precision.

Stat De-Id (Uzuner et al., 2008; Sibanda, 2006) takes a machine learning approach using Support Vector Machines (SVM) as the learning algorithm. Features cover aspects of the target word as well as of the immediate context of the target. Conditional Random Fields (CRF) (Lafferty et al., 2001) are being used increasingly in de-identification research. Two examples are Health Information DE-identification (HIDE) (Gardner and Xiong, 2008) and the Mitre Identification Scrubber Toolkit (MIST) (Aberdeen et al., 2010; Deleger et al., 2013). Several of these de-identification systems (see also Douglass et al. (2004) and Neamatullah et al. (2008)) show excellent results rivaling manual de-identification. While most de-identification systems score well in terms of recall, they do produce quite a large amount of false positives (see Ferrández et al. (2012a)). This compromises the usability of the de-identified documents, as medically relevant data may have been removed.

In this paper, we present a de-identification case study following best practices from the state of the art. A machine learning approach is taken, using features based on dictionaries and string and pattern matching techniques. The objective of this study is to develop a de-identification system for clinical notes in Dutch, a language for which de-identification training data are not available. We evaluate three machine learning setups on a training set of 100 manually annotated medical notes and test the best performing setup on 100 previously unseen medical notes, the performance of which is manually evaluated at the hospital site.

2 Methods

2.1 Data set

The training set consists of 100 documents randomly selected from the Antwerp University Hospital (UZA) EHR system. This data set consists of (discharge) letters, comprising 52,829 words in total. These words have been annotated manually according to the following Personal Health Information (PHI) classes: *Name*, *Date*, *Address*, *ID* (indicating a personal identification code such as a social security number), and *Hospital*. 2,968 words were manually marked as containing PHI. Their occurrence rates are shown in Figure 1.

Figure 1: Average frequency per PHI class over total number of words (n=52.829)

For privacy reasons, all PHI words in these documents have been obfuscated in the hospital before we obtained them. The PHI-labelled documents were then reconstructed with fictitious names, addresses, etc. to enable their use as a training set. A test set with 100 randomly selected documents was held internally at the Antwerp University Hospital (UZA) and was manually annotated to be used for later testing (see Section 3.3). However, training and test set differ in quality since the former was manually obfuscated after manual de-identification to protect patient privacy and the latter was unaltered.

2.2 Experimental setup

For the development of our de-identification system using the training set described above, we apply the following experimental setup. First of all, the texts in the dataset are tokenized (i.e. splitting the text in individual words and removing punctuation). Next, features are derived and calculated. In a third step, the resulting set of derived features with associated PHI class per word is used for training. In a set of experiments, we (1) assess the performance of the classifiers for the individual PHI classes, (2) evaluate how adding more training data affects performance, and (3) validate the performance on 100 previously unseen documents. Due to the high cost of manual annotation, our training set is rather small. As a result, the performance scores can only be interpreted as indicative of performance in a realistic environment.

All results in Experiments 1 and 2 are averaged over fifty independent runs, each selecting different sets of training and test sets from the original training set. In each run, ten random documents are withheld as test set. In Experiment 1, the remaining ninety are used for training, while in Experiment 2, a learning curve is constructed, showing the effect of a stepwise (step size=10) increase in training set size.

2.3 Feature engineering

Since the choice of features affects algorithm behavior and performance (Kim et al., 2011; Sibanda, 2006), selecting features that discriminate PHI from non-PHI and are able to indicate the differences between the various PHI classes (Gardner and Xiong, 2008) is crucial. Because of the limited data available for training, external dictionaries are indispensable.

We use four types of features: (i) direct target word characteristics, (ii) pattern matching features, (iii) dictionary features, and (iv) contextual word features. Direct target word characteristics indicate the presence of capitalisation, punctuation, and numbers and includes word length information. Pattern matching features are linked to regular expressions that refer to social security numbers or date patterns. Dictionary features indicate whether the target word is present in a PHI dictionary (i.c. dictionaries of first and last names, streets, cities, hospital names, healthcare institutions, salutations) or whether it is part of a word group that is present in a PHI dictionary. For word groups, we take into account a context of three words to the right of the target word (i.e. sliding window size=4). For computational efficiency, we use a suffix tree algorithm by Ukkonen (1995). Contextual word features indicate whether words in the immediate context (i.e. left context=3, right context=3; sliding window size=7) of the target word have characteristics that might influence the classification of the target word (e.g. punctuation, capitalisation).

2.4 Classification

We apply three classification setups, each offering their own advantages for different data sets, dependent on the data set size, the heterogeneity of the data set, and the total number of classes. We use Weka (Witten and Frank, 2005), a toolkit for machine learning, for classification with Random Forests. For Support Vector Machines (SVM), we use the libSVM (Chang and Lin, 2011) library. In future de-identification experiments, we will evaluate Conditional Random Fields as well.

SVMs calculate an optimal decision boundary between two classes (Chang and Lin, 2011), are powerful with high-dimensional data and promote the use of local context features. For de-identification with several PHI classes, multi-class classification is required. We test (i) a one-versus-one learning scheme (cf. 'OOSVM'), where the binary classifiers distinguish between each pair of classes and (ii) a one-versus-all scheme (cf. 'OMSVM'), where each class is distinguished from the other classes simultaneously. Both schemes apply majority voting with equal weights assigned to each (PHI as well as non-PHI) class.

Random Forests is a machine learning technique that generates multiple random Decision Trees (Breiman, 2001). Each of these trees randomly selects features and assigns a particular class to each instance containing those features. A voting system decides which of these decisions is finally assigned, potentially leading to a more robust decision since it is supported by multiple trees. The total number of trees is customisable, but a high number of trees increases training time. We tested multiple numbers of trees, but selecting ten random trees (cf. RF10) yielded the best balance between precision, recall, and training time.

2.5 Evaluation measures

We present results in terms of precision, recall, and F-score. We consider recall to be the most important measure for de-identification as it shows the number of PHI-items actually retrieved by the algorithm divided by the number of PHI items present. Precision indicates how many of the PHI items identified are actually correct. F-score is calculated as the harmonic mean between precision and recall. Precision and recall are macro-averaged, in a way that all classes have an equal weight in the end result.

3 Results

We present results of three experiments: we (1) evaluate the performance of the proposed method for five PHI classes, (2) perform a learning curve experiment to investigate how performance is affected by increasing training set size, and (3) evaluate the best experimental setup on a previously unseen test set of 100 documents.

3.1 Performance on individual PHI classes

Recall and precision are very similar for most classes, as is shown in Table 1, except for the *Name* and *ID* classes. This can be explained by the wide variety in types of IDs and the larger ambiguity between names and non-PHI words (e.g. 'Vrijdag', the Dutch word for 'Friday', also represents a last name found in libraries with a relatively high

	Name	*Date*	*Address*	*ID*	*Hospital*
OOSVM					
Recall	91.2	95.9	95.6	79.9	95.0
Precision	88.6	98.0	98.2	95.3	98.6
F-score	90.1	96.9	96.8	86.9	96.8
OMSVM					
Recall	91.2	95.8	96.2	77.2	95.4
Precision	88.9	98.0	98.4	95.3	98.6
F-score	90.0	96.8	97.3	85.3	97.0
RF10					
Recall	87.4	95.0	92.5	75.8	75.3
Precision	95.1	98.4	98.5	99.4	97.8
F-score	91.1	96.6	95.4	86.0	85.1

Table 1: Results per PHI class and classification setup

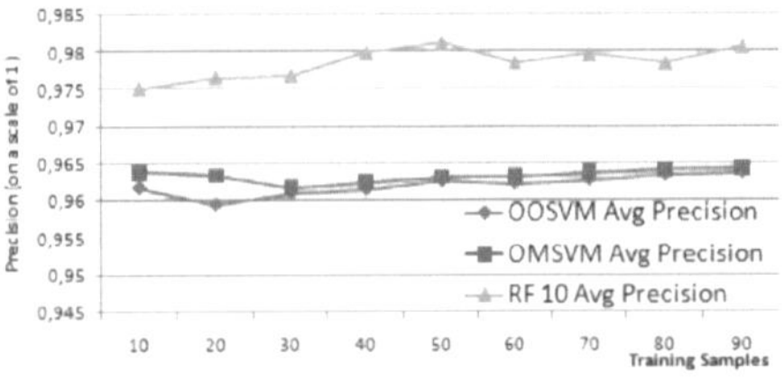

Figure 2: Average precision per setup with 90 training and 10 test documents

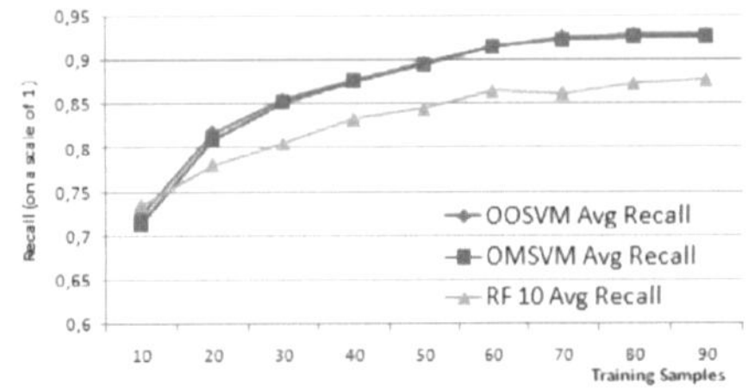

Figure 3: Average recall per setup with 90 training and 10 test documents

frequency). It should be noted that performance is calculated per word in a (potentially multi-word) name. If only part of the name gets classified as a *Name*, it is counted as a false negative, although the largest part of the name will be removed from the text.

Overall, our SVM setups show a higher recall and F-measure than the Random Forest setup, while the latter has a higher precision. With 90 training documents, an average F-measure of 91.5% for the Random Forest method and an average F-measure of 94.5% for the one-against-one SVM setup is achieved.

3.2 Learning curve

To assess the amount of manually annotated data required, we increase the number of training documents in a learning curve experiment. Figures 2 and 3 represent precision and recall scores with varying training set size. The RF10 method has a generally higher precision than the SVM setups, but also a lower recall. Precision remains relatively constant for all methods and recall values seem to converge asymptotically.

3.3 Results on a previously unseen test set

In this experiment, we evaluate the algorithm in a more realistic setting where the algorithm - built from a limited set of manually obfuscated training data (cf. Section 2.1) - is tested on previously unseen test data and evaluated by a hospital staff member. The test data are qualitatively different since they were not subject to obfuscation. Experiments were conducted with OOSVM - the machine learning setup that yielded the best performance in the experiments described above - using 100 obfuscated documents for training and yielded a recall of 89.12% and a precision of 93%, which is lower than the performance on the *obfuscated* test data in Section 3.2.

Error analysis revealed that the use of all-caps (first and last) names and addresses is widespread in the test documents, whereas the training data were manually obfuscated and contained no all-caps names and addresses. Since capitalisation is a feature (cf. Section 2.3) in our de-identification system, the difference in quality between training and test data can explain the drop in performance.

3.4 Time measurements

Time measurements have been taken to check whether the de-identification algorithm is applicable to a larger set of documents. A de-identification speed of 109 ms/document (assuming an average length of 500 words) was achieved when de-identifying with the OOSVM method, while the Random Forest method only needed 42 ms/document. The OMSVM method requires a de-identification time of 205 ms/document.

If OOSVM, the best performing setup, would be used to de-identify documents from the hospitals EHR system on a daily basis, the processing time would be a matter of minutes. The larger amount of time needed to use the one-against-one SVMs rather than the Random Forest method is worth it, since the performance of the former is significantly better.

4 Discussion

The results suggest that the de-identification algorithm we developed achieves reasonable performance considering the limited set of training data it is based on. However, to be of practical use without manual confirmation, de-identification recall should be as high as possible, making sure that no PHI remains in the text. High precision is of secondary importance, as long as the algorithm does not identify too many non-PHI words as containing Personal Health information, which can cause medically relevant information to be lost during de-identification.

The learning curve experiments show that recall scores start to converge asymptotically, which may indicate that relatively small amounts of training data already yield fair results, while the increase in precision with increasing training set size seems limited. However, we are aware that the data set is too limited to draw conclusions from these results.

The test on a non-obfuscated, previously unseen test set indicates that minor feature improvements and a more representative training set are needed. Although the current approach with a previously manually obfuscated training set is non-scalable, it allows us to automatically create a more representative training set from another dataset.

The results of the Random Forest method can be improved when increasing the amount of trees. However, this also increases training time linearly, with a minimal increase in performance. Recall scores of the current Random Forests setup are insufficient for most PHI classes.

5 Conclusion

In this paper, we presented a machine learning approach to de-identification based on a limited set of manually annotated Dutch-language clinical notes. We compared three types of classification approaches and found the one-versus-one SVM setup to be the method of choice for this particular case study. In terms of recall - which we consider the most crucial evaluation measure for practically usable de-identification - it is better than the Random Forest classifier, which in its turn scores better in terms of de-identification time and precision. Learning curve results seem to indicate that the amount of training data needed converges to an asymptote quite early in the curve.

We plan several extensions to the algorithm: adding syntactic (e.g. part-of-speech tags) and semantic features, investigating the use of semi-supervised learning to automatically increasing the set of training data, and testing Conditional Random Fields for classification. Another next step is the expansion to an ensemble method for two of our classifiers, taking advantage of properties of both classifiers.

6 Acknowledgments

The authors would like to thank Antwerp University Hospital (UZA) - in particular the innovative ICT programme - University of Antwerp, and Artesis Hogeschool Antwerpen for their support.

References

J. Aberdeen, S. Bayer, R. Yeniterzi, B. Wellner, C. Clark, D. Hanauer, B. Malin, and L. Hirschman. 2010. The mitre identification scrubber toolkit: design, training, and assessment. *International journal of medical informatics*, 79(12):849–859.

Belgian Data Protection Act. 1993. Consolidated text of the Belgian law of December 8, 1992 on Privacy Protection in relation to the Processing of Personal Data as modified by the law of December 11, 1998 implementing Directive 95/46/EC.

L. Breiman. 2001. Random forests. *Machine learning*, 45(1):5–32.

C. Chang and C. Lin. 2011. Libsvm: a library for support vector machines. *ACM Transactions on Intelligent Systems and Technology (TIST)*, 2(3):27.

L. Deleger, K. Molnar, F.i Savova, G.and Xia, T. Lingren, Q. Li, K. Marsolo, A. Jegga, M. Kaiser, and L. Stoutenborough. 2013. Large-scale evaluation of automated clinical note de-identification and its impact on information extraction. *Journal of the American Medical Informatics Association*, 20(1):84–94.

M. Douglass, G.D. Clifford, A. Reisner, G.B. Moody, and R.G. Mark. 2004. Computer-assisted de-identification of free text in the mimic ii database. *Computers in Cardiology*, 29:641–644.

European Data Protection Directive. 1995. Directive 95/46/EC of the European Parliament and of the Council of 24 October 1995 on the protection of individuals with regard to the processing of personal data and on the free movement of such data.

O. Ferrández, B.R. South, S. Shen, F.J. Friedlin, M.H. Samore, and S.M. Meystre. 2012a. Generalizability and comparison of automatic clinical text de-identification methods and resources. In *Proceedings of the AMIA Annual Symposium*, pages 199–208.

O. Ferrández, B.R. South, S. Shen, and S. Meystre. 2012b. A hybrid stepwise approach for de-identifying person names in clinical documents. In *Proceedings of the 2012 Workshop on Biomedical Natural Language Processing*, pages 65–72. Association for Computational Linguistics.

J. Gardner and L. Xiong. 2008. HIDE: An Integrated System for Health Information DE-identification. In *Proceedings of the 21st IEEE International Symposium on Computer-Based Medical Systems*, pages 254–259.

HIPAA. 1996. Health Insurance Portability and Accountability Act of 1996.

Y. Kim, E. Riloff, and S.M. Meystre. 2011. Improving Classification of Medical Assertions in Clinical Notes. In *Proceedings of the 49th Annual Meeting of the Association for Computational Linguistics: Human Language Technologies: Short Papers*, volume 2, pages 311–316.

J. Lafferty, A. McCallum, and F. Pereira. 2001. Conditional random fields: Probabilistic models for segmenting and labeling sequence data. *Proceedings of the Eighteenth International Conference on Machine Learning*, pages 282–289.

S. Meystre, F. Friedlin, B. South, S. Shen, and M. Samore. 2010. Automatic de-identification of textual documents in the electronic health record: a review of recent research. *BMC medical research methodology*, 10:70:1–70:16.

I. Neamatullah, M.M. Douglass, L.H. Lehman, A. Reisner, M. Villarroel, W.J. Long, P. Szolovits, G.B. Moody, R.G. Mark, and G.D. Clifford. 2008. Automated De-Identification of Free-Text Medical Records. *BMC Medical Infomatics and Decision Making*, 8(32):1–17.

Privacy Rule. 2002. Standards for Privacy of Individually Identifiable Health Information: Final Rule. *Federal Register 53181*, 67(157):53181–53273. (codified at 45 CFR 160 and 164).

T.C. Sibanda. 2006. Was the Patient Cured? Understanding Semantic Categories and Their Relationships in Patient Records. Master's thesis, Department of Electrical Engineering and Computer Science, Massachusetts Institute of Technology.

L. Sweeney. 1996. Replacing Personally-Identifying Information in Medical Records, the Scrub System. In *Proceedings of the AMIA Annual Fall Symposium*, pages 333–337.

E. Ukkonen. 1995. On-line construction of suffix trees. *Algorithmica*, 14(3):249–260.

Ö. Uzuner, T.C. Sibanda, Y. Luo, and P. Szolovits. 2008. A de-identifier for medical discharge summaries. *Artificial intelligence in medicine*, 42(1):13–35.

I.H. Witten and E. Frank. 2005. *Data Mining: Practical Machine Learning Tools and Techniques*. Morgan Kaufmann, San Francisco, 2 edition.

Helping parents to understand rare diseases

Marina Sokolova
CHEO Research Institute and University of Ottawa
sokolova@uottawa.ca

Ilya Ioshikhes
University of Ottawa
iioschik@uottawa.ca

Hamid Poursepanj
University of Ottawa
Hpour099@uottawa.ca

Alex MacKenzie
CHEO Research Institute and University of Ottawa
mackenzie@cheo.on.ca

Abstract

Rare diseases are not that rare: worldwide, one in 12-17 people will be affected by a rare disease. Newborn screening for rare diseases has been adopted by many European and North American jurisdictions. The results of genetic testing are given to millions of families and children's guardians who often turn to the Internet to find more information about the disease. We found 42 medical forums and blogs where parents and other related adults form virtual communities to discuss the disease diagnosis, share related knowledge and seek moral support. Many people (up to 75% in some population groups) look for professional medical publications to find reliable information. How can it be made easier for these non-medical professionals to understand such texts? We suggest that recommender systems, installed on web sites of research and teaching health care organizations, can be a tool that helps parents to navigate a massive amount of available medical information. In this paper, we discuss NLP architecture of such a system. We concentrate on processing epistemic modal expressions and helping the general public to evaluate the certainty of an event.

1 Introduction

A rare disease is identified as a life-threatening or chronically debilitating condition affecting not more than 5 in 10.000 persons (Cornel et al., 2013). Accumulatively, rare diseases are not uncommon. In UK, one in 17 people will be affected at some point in their lives; this equates to more than 3.5 million people. Most of these will be children, and 30% of individuals with a rare disease will die before they are five years old. [1] In Canada, one in 12 people suffer from a rare disease, and the number of identified rare diseases identified constantly increases.[2]

A prevailing understanding is that early diagnosis and treatment may ease the burden of a disease. Thus, newborn screening for rare diseases has become a routine procedure in USA, Canada and the member states of the European Union. Nevertheless, Raffle and Gray (2007) note that screening programs can induce harm. Whereas the programs improve health status in patients by diagnosing them early and treating optimally, after the screening, parents and guardians of the newborns receive a substantial amount of new information about health of their children. The results may cause parental stress and anxiety, among other negative factors. To aid in navigation through the screening results, health care authorities organized web-based resource centers, such as a joint web site by the Newborn Screening Ontario program and Children's Hospital of Eastern Ontario[3] or the Newborn Screening web site by the National Health Service[4].

Emergence of user-friendly online technologies prompted the general public to turn to the Internet to gain more knowledge on health-related issues, a phenomenon often referred to as Dr. Google. A 2011 survey of the US population estimated that i) 59% of all adults have looked online for information about health topics such

[1] http://blogs.biomedcentral.com/bmcblog/2013/02/28/what-is-the-cost-of-rare-diseases/
[2] http://rare-diseases.ca/
[3] http://www.newbornscreening.on.ca/bins/index.asp
[4] http://www.nhs.uk/Livewell/Screening/Pages/Newbornscreening.aspx

as a specific disease or treatment, ii) 18% of adults have consulted online reviews of particular drugs or medical treatments, iii) 29% of all adults sought online health information related to somebody's else medical condition (Fox 2011; Fox 2011a). Preference in online searches related to specific health problems was previously reported by Nicholas et al. (2003).

At the same time, the general public does not consider different sources of the available health information as being equal. 75% of non-medical professionals aim their online searchers at professional medical sites and publications (McMullan, 2006). Many individuals prefer to have an access to a complex and complete information (80%), whereas some feel that the information usually accessed is too basic (45%) (ibid).
An important question arises as to how well the readers can understand the information they retrieve. Eysenbach (2003) reported that patients who sought health information online felt that Internet information can be overwhelming (31%), conflicting (76%) and confusing (27%).

The system that we are building aims to help individuals, specifically parents of newborns, to navigate and assess medical publications about rare diseases.

2 Motivation

It has been documented that many parental users of the Internet are first-time parents (Oprescu et al, 2013) and parents with young children (Balka and Butt, 2006). They might not have a hands-on experience or practice in dealing with a complex medical issue and information related to it and can be easily overwhelmed. This is especially true for parents of newborn babies which were screened positive for one of the rare diseases.

The screening happens at the very beginning of the child's life. A positive result may cause parents' insecurity and increase their uncertainty about future (Brashers, 2001). Gaining new knowledge through relevant information is one of the tools that can alleviate stress and anxiety (Brashers, 2000). It is natural that the affected parents search for information about the diagnosis and turn their attention to professional medical publications.

In the quest for knowledge, parents face a challenge of understanding a complex text that includes assessment of the relevance of the retrieved information and ability to differentiate among available options. This necessitates, among other required skills, ability to discriminate between different degrees of certainty and evaluate the likelihood score found in the text (Holmes, 1982; Morante and Sporleder, 2012).

3 Medical Forum Data

To get a notion of the public concerns and questions, we automatically extracted and manually analyzed messages posted on medical forums and blogs dedicated to newborn screening for rare diseases.

We selected 42 forums frequently visited by parents and families concerned about newborn screening and its consequences (e.g., forums.familyeducation.com,www.justmommies.com/babies/newborn,www.parentingforums.org/forum.php). We did not require research ethics review for this study as all of the data collected and used was from publically available sources. Nevertheless, we confirmed with our institutional research ethics board that no review of research on public data sets was necessary.

To find what parents think and discuss, we followed two strategies to collect data: a manual search and automated crawling of parenting forums and blogs. In our manual data collection, we used Google to find blogs, comments and medical forums that could post messages with the relevant content. Our queries were built from all the possible combinations of the following three sets of phrases:

- {Forum, Bulletin Board, Message Board, BBS, Threads}
- {Newborn, infant}
- {Genetic Predisposition Testing, Genetic Screening, Predictive Genetic Testing, Genome Sequencing}

For instance, the query {newborn genetic screening threads} is used to search for the related forums. After finding relevant blogs or threads, we downloaded the comments.

To download the comments we used Selenium API[5], Hibernate API[6], MySQL database, and Java programming language. In automatic data collection, we used Apache Nutc crawler application[7] in combination with Apache Solr search platform[8] to crawl handpicked forums.

First, we performed manual search to find forums with the related phrases mentioned above. Then we used URLs of the selected forums as a seed for Nutch. We stored the collected HTML pages in MySQL database. Next, we searched these pages with Apache Solr, which uses the cosine similarity metric to find related page to the search query:

$$\text{similarity} = \cos(\theta) = \frac{A \cdot B}{\|A\|\|B\|} = \frac{\sum_{i=1}^{n} A_i \times B_i}{\sqrt{\sum_{i=1}^{n} (A_i)^2} \times \sqrt{\sum_{i=1}^{n} (B_i)^2}}$$

where A and B are the term frequency vectors of the query and a document. We marked page as relevant if it contained any three-phrase combination built from three sets of phrases mentioned earlier in this section.

After finding pages with relevant comments, their content was downloaded and analyzed independently by two authors. We found that parents and other involved individuals are often concerned about the following issues:

- Prevalence and severity of the disease
- Available treatment and the effects of an early treatment
- A possible course of the disease
- Reliable tests
- Health care facilities
- Medications and their side effects
- Future use of the results

Table 1 lists some examples of messages. We keep all the original spelling, punctuation, and grammar.

[5] http://docs.seleniumhq.org/, accessed: July. 2, 2013
[6] http://www.hibernate.org/, accessed: July. 2, 2013

[7] http://nutch.apache.org/#What+is+Apache+Nutch%3F, accessed: July. 2, 2013
[8] http://lucene.apache.org/solr/ , accessed: July. 2, 2013

Table 1: Online messages on newborn screening

Concern	Message
Prevalence and severity	I was 39 when my son was born. 1:810 for downs and 1:10,000 for tri 18/13. The risk for ANY chromosomal issue is 1:80 for someone age 39.
Available treatment and the effects of an early treatment	she said if he starts to get congestion then do a breathing treatment and call next day if he is not doing better. I was confused about this as well as I thought he would need those treatments immediately....
A possible course of the disease	nothing is written in stone for CF these days. Docs told my parents I wouldn't live past 21. I'm 27yo, working on a PhD, and this past spring I finished my second Half-Ironman Triathlon.
Reliable tests	I'm concerned because of the number of false-positives I read about, and further testing to eliminate that worry have a small chance of causing death to a perfectly healthy baby.
Health care facilities	best to do them in a cf[9] center cause it all depends on the lab you do the test in.
Medications and their side effects	I have a 6 month old son that was born deaf. He just recieved a baha hearing aid, and it has done some good, but not as much as we had hoped for
Future use of the results	And you have no problem with this government owning their genetic code, potentially knowing illness, disabilities, strengths, weaknesses and potential? A trusting soul you are indeed.

[9] Cystic fibrosis - an autosomal recessive genetic disorder that affects most critically the lungs, and also the pancreas, liver, and intestine.

4 Epistemic expressions

Our current project focuses on disambiguating epistemic modal expressions. Previously, several studies connected the use of modal verbs (can, might, should), amplifiers (certainly, definitely) with the level of expectation of an event (Henriksson and Velupillai, 2010; Sokolova and Lapalme, 2011).

However, these studies did not categorize epistemic expressions based on their strength of conviction in the event happening. We categorize the strength of conviction by assigning six categories: *impossible*, *improbable*, *uncertain*, *possible*, *probable*, and *certain* (Horn, 1989; Sokolova et al, 2010). Table 2 shows the categories and expressions.

Table 2: Examples of epistemic expressions

Epistemic category	Expressions
Impossible	Never happens
Improbable	Hardly expected
Uncertain	We unsure
Possible	Perhaps
Probable	There is a certain risk
Certain	We always see

The suggested six categories add three negative categories (*impossible*, *improbable*, *uncertain*) to common positive happenstance categories (*possible*, *probable*, *certain*) (R. Saurı´ and J. Pustejovsky, 2009).

Language expressions corresponding to the categories contain extensional modifiers, i.e. modifiers of degree and happenstance (Sokolova and Lapalme, 2011). The modifiers can be modal verbs (can, must, would), adverbs (likely, mostly), adjectives (common, rare) and quantifying pronouns (every, none). These expressions should be disambiguated in the context of a sentence or a clause.

Below we categorize a few expressions found in articles on rare diseases:

- Impossible: such off-target gene modulating effects are currently impossible to predict
- Improbable: in part out of concern that recruitment of eligible subjects with SMA Type I would be difficult because of their high level of inter-current illness and mortality in childhood
- Uncertain: these studies could not rule out an additional contribution resulting from restoration of SMN levels in muscle
- Possible: raising the possibility that intrinsic responses to low levels of SMN in skeletal muscle may also contribute directly to SMA pathogenesis
- Probable: Studies have shown that restoring SMN protein levels in neurons can significantly ameliorate disease progression
- Certain: A neighboring nearly-identical copy of this gene, SMN2, is invariably present in individuals with SMA

Our next step was to assess the presence of epistemic expressions in articles on rare diseases.

5 Empirical evidence

For our preliminary study we decided on a group of articles dedicated to spinal muscular atrophy (SMA), a neurodegenerative disease affecting 1 in 11000 newborns world- wide (Farooq et al, 2013). We selected six full-length research articles. The articles were published in Orphanet Journal of Rare Diseases, Neurodegenerative Diseases (an open source book), Journal of Clinical Investigation, Plos One, and Human Molecular Genetics (2 articles).

The articles' content covers most of parent concerns (see Table 1): a) severity of the disease, b) available treatment and the effects of an early treatment, c) a possible course of the disease, reliability of tests, d) medications. Thus, the article selection allows the empirical results to be representative of the information parents will find. At the same time, we cover language expressions belonging to different authors.

Three articles were written by a team of leading researchers in SMA; these authors have written papers that receive a high rank in SMA search through Google Scholar; hence, there is a high probability that parents looking for the SMA information will first encounter papers published by this team (Articles A). Three other articles were written by other teams working on SMA research (Articles B). We report the descriptive statistics in Table 3; *vocabulary* signifies different words in the text.

Table 3: Vocabulary richness of the six articles; *d.leg.* – vocabulary with occur. = 1, *h.leg.* – vocabulary with occur. = 2, occ > 5 – vocabulary that occur 6 and more times.

Articles A					
#	**words**	**vocab.**	***d. leg.***	***h. leg.***	**Occ >5**
1a	6894	1661	1008	254	159
2a	4492	1374	827	242	121
3a	6629	1506	871	246	185
Articles B					
#	**words**	**vocab.**	***d. leg.***	***h. leg.***	**Occ >5**
1b	10731	2658	1344	441	317
2b	6009	1557	896	246	173
3b	5094	1125	598	191	157

We looked for epistemic expressions related to the main concerns found on medical forums (see Table 1). For the information retrieval, we built N-gram models of each article (N = 1, …, 4). We tokenized data by splitting along spaces and punctuation marks. We kept the original capitalization to preserve the beginning of sentences. We used combinations of the seed words to find N-grams with the epistemic meaning. Table 4 lists examples of the seed words.

Table 4: Examples of words used in search of epistemic expressions.

Part-of-speech	Examples
Adverbs	Possibly, perhaps
Adjectives	Impossible
Modal verbs	Can, may, should
Negations	No, not
Nouns	absence, presence
Quantifying pronouns	Every, none

To avoid counting the same expression multiple times, we filtered out those bi-grams which first word overlaps with the second word of another epistemic bi-gram. From the list of tri-grams, we filtered out those which first word overlaps with the third word of another epistemic tri-gram. For instance, we filtered out not be as a possible extension of may not.

Manual analysis showed that the most frequent bi- and tri-grams expressed negative and positive conviction, supporting the proposed expansion of epistemic categories by negative *impossible, improbable* and *uncertain.* Table 5 lists the most frequent epistemic bi- and tri-grams found in the articles.

Table 5: Five most frequent not overlapping epistemic bi- and tri-grams per article.

Articles A		
Article1a	Bi-grams	may not, may be, can be, will be, where possible
	Tri-grams	may not be, the hope is, would have a, majority of these, and where possible
Article 2a	Bi-grams	which can, SMA can, and can, must be, may have
	Tri-grams	which can cross, no cure for, SMA can be, SMA is primarily, which may have
Article 3a	Bi-grams	potential treatment, such promising, promise for, of approximately, suggest that
	Tri-grams	potential therapeutic compounds, potential treatment strategy, such promising agent, found to be, There could be
Articles B		
Article 1b	Bi-grams	can be, able to, will be, would be, could be
	Tri-grams	the potential to, the false discovery, can be used, may not be, there were no
Article 2b	Bi-grams	absence of, implicated in, as expected, potential to, the possibility
	Tri-grams	absence of any, confirmed in a, raising the possibility, potential to act, supported by several
Article 3b	Bi-grams	a putative, probably carry, could be, should be, would be
	Tri-grams	result not shown, putative gene conversion, affected children could, observations should be, This change would

We hypothesized that the use of the expressions closely relates to the content of the article. For example, articles reporting on clinical trials, treatments, medications may have a high frequency of epistemic expressions as due to necessity of drawing conclusions and implications for future patients. We list the topics of the six papers and frequencies of the top epistemic bi- and tri-grams in Table 6.

Table 6: Article topics and frequencies of the epistemic expressions (x10^{-3})

Articles A			
#	Topics	bigrams	trigrams
1a	disease therapy, preclinical drug development, generalizable screening methods	22.4	14.0
2a	Classification, diagnosis, background for SMA	5.0	2.9
3a	PRL treatment in mice, potential therapeutic compounds	2.1	1.6
1b	Identification of novel candidate biomarkers associated with disease severity in SMA	3.2	1.1
2b	intrinsic pathology of skeletal muscle, novel biomarkers in SMA	6.3	2.0
3b	molecular analysis of the SMN and NAIP Genes	4.3	2.3

Looking at the topics of the papers, we can see that preclinical drug development corresponds to the largest frequency of the epistemic expressions in the text.

6 Parent Advisor

In the medical domain, uncertainty and misunderstanding of information can imperil lives and incur significant costs on health care systems (McCoy et al, 2012). Although a thirst for medical knowledge among non-medical readers has been documented (see Section 1), there are not many developed NLP tools and methods that help such readers to understand medical texts.

Although epistemic disambiguation is important for text understanding, other text analysis tasks are essential in order to build an effective system that helps parents to understand medical publications related to rare diseases. Hence, the proposed system name is the Parent Advisor.

We suggest that such the Parent Advisor is organized as a pipeline of NLP and Text Mining tools, each serving a special purpose (Figure 1):

1. social media analyzer, to identify concept shift in parents' concerns;
2. article content classifier, to select relevant articles for further analysis;
3. relevant article ranker, to evaluate the usefulness of the selected articles;
4. factual information extractor, to retrieve information related to parents' concerns
5. epistemic disambiguator, to assess the retrieved factual information.
6. the output ranker, which ranks the factual information according to the assigned epistemic categories.

We also suggest that human participation should be incorporated in the system functioning. To support the actuality of text analysis, parents can be surveyed and polled either on a regular basis or in relation to medical and health-related events (e.g., a new discovery, a proposed change in health care) (Fox, 2011). Medical professionals should be involved in the article ranking, to ensure the quality of the selected publications. Additionally to standard text annotation, a team of communication and medical professionals can assist in the factual information extraction and epistemic disambiguation (Scott et al, 2012).

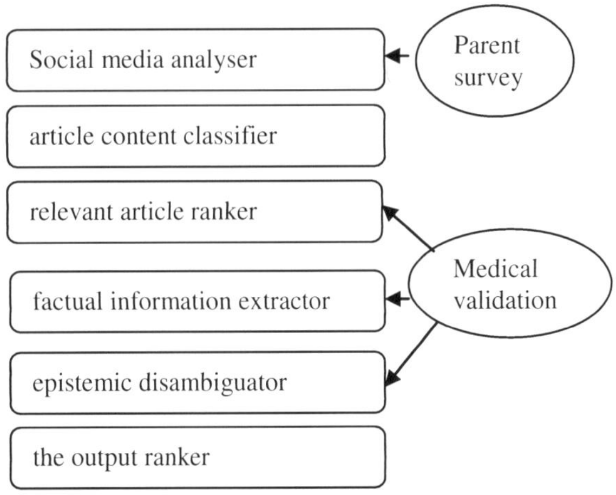

Figure 1: the Parent Advisor system.

For a given query, the system will release the information ranked accordingly to the epistemic categories. For example, the information deemed certain will be ranked higher than the information deemed probable or possible (Table 7). When releasing the information marked with negative categories, we plan to label it with "Caution".

Table 7: A possible output for the query " orphan diseases, protein, therapy".

#	Extracted text	Annotation (not shown to parents)
1	Pharmacological chaperones stabilize the folding of mutant proteins and allow for correct trafficking of the enzyme	*Certain*
2	mRNA serves as a valid proxy for protein level more often than not.	*Probable*
Caution	[the pharmacologic upregulation of gene activity and mRNA level] will not work if the mutated protein has any dominant negative effect.	*Impossible*
Caution	even if a protein:RNA correlation is observed in vitro, a given transcript response detected in cell culture may not hold true for a whole organism.	*Improbable*

7 Related Work

The semantic analysis of biomedical and clinical texts mainly focuses on identification and disambiguation of medical terms and events (Cohen et al, 2011; Demner-Fushman et al, 2010; Savova et al, 2011) including temporal characteristics of events (Boytcheva et al, 2012). Emergence of electronic health records enabled studies of epistemic expressions, often linking them with the diagnosis of a patient (McCoy et al, 2012; S. Velupillai' 2010). Expressions of certainty in the medical publications, however, are not well studied, although it is natural to expect a medical publication to contain epistemic expressions: while presenting factual information, a publication also conveys the authors' inference from the facts and conviction in the event happenstance.

Categorizing text into speculative and non-speculative parts partially addresses the problem (Szarvas, 2008; Sanchez et al, 2010) as such division only differentiates the *certain* (aka non-speculative) category from other epistemic categories. We, however, want to analyze language expressions of several epistemic categories.

Vagueness in clinical records was studied by Emanuel and Emanuel (1989) and more recently by Hyland (2006) and Scott et al (2012). These studies concentrate on the use of happenstance modifiers (Sokolova and Lapalme, 2011), also known as linguistic hedges (e.g., *possible, probably, few, consistent with*). The goal is to develop an automated analysis of diagnosis and symptom information extracted from electronic patient records (Scott et al, 2012). Although the task is similar to our goal, we plan to work with information extracted from medical publications.

A more complex approach would be to introduce a pragmatic component into the epistemic analysis of medical publications. This approach differentiates between a hypothesis, accepted knowledge, and new experimental knowledge (Nawaz et al, 2010). Such categorization may be useful in the recommender system designed for the general public. In future, we plan to work on the hypothesis - experimental knowledge division.

Our work also closely relates to text understanding and interpretation (Bos, 2011). With the development of Internet search engines, text understanding and interpretation mainly focused on retrieval of texts relevant to the query. A few systems develop a more advanced and deep text exploration which interprets text on demand from the system users (Dunne et al, 2012). The systems are field-specific, often built on ontology, and are designed to help professionals working in the field (Dunne et al, 2012; Wimalasuriya and Dou, 2010). An advanced, semantically based information retrieval is performed by question-answering systems. In medicine, such systems assist clinicians to find clinically-relevant information (Cao et al, 2011). Our goal, in contrast, is to build a system that helps the general public to understand professional medical text.

Note that during the related work analysis, we could not find published studies that relate parent concerns to social media to rare disease information.

8 Future Work

Our immediate future work will focus on epistemic annotation of a large collection of SMA articles. What can be considered a sufficient size of the annotated corpus is an open question in BioNLP: the physical chemistry and biochemistry Core Scientific Concepts corpus has 265 articles (Liakata, 2010), the bioinformatics' Bioscope corpus has 9 articles and 1273 abstracts (Vincze et al, 2008).

We suggest that the number of annotated articles can be linked to the annual rate of SMA publications. For example, the Google Scholar search for "spinal muscular atrophy"[10] retrieved 278 articles[11] published in 2012 – 2013. A similar PubMed search[12] retrieved 222 articles[13]. Hence, we aim to annotate 150 - 200 SMA articles and abstracts.

We want the article profiles be representative of the concerns expressed on the medical forums (listed in Table 1). For example, after adding the term "treatment" to both searchers we retrieved 77 articles through Google Scholar and 99 articles through PubMed, while substituting "treatment" by "drugs" we retrieved 45 articles and 7 articles respectively. Thus, our annotated data should include 60-80 treatment-oriented and 10-30 drug-oriented articles and abstracts.

The article and abstract selection is another open question in the article annotation. Our criterion is the corpus compliance with the expected retrieval results of the Web search. Hence, for each topic, we will select articles based on their relevance.

To ensure that the corpus is consistent with the public concerns, we will continue to analyze medical forums in order to keep updates on the general public response on newborn screening for rare diseases.

9 Conclusions

We presented a preliminary work on the system which we call Parent Advisor. Parent Advisor can help parents to understand medical publications related to rare diseases. The topic of rare diseases became a subject of many discussions, as more and more jurisdictions adopted a policy of newborn screening for rare diseases.

To make our future system actual and useful for parents, we have studied messages posted on 42 medical forums. These forums are frequently visited by parents and families who have questions related to newborn screening for rare diseases. We identified several issues that concern the general public most (prevalence and severity of the disease, available treatment and the effects of an early treatment, a possible course of the disease, reliable tests, health care facilities, medications and their side effects, future use of the results).

We chose epistemic disambiguation as the focus of our first sub-project in building Parent Advisor. We identified six epistemic categories (*impossible, improbable, uncertain, possible, probable, certain*), thus expanding a usual range of positive categories (*possible, probable certain*) by negative categories (*impossible, improbable, uncertain*).

In this paper, we also outlined the architecture of the system, sketched human-system collaboration desirable for the system to be effective, and presented a detailed plan for future work.

Acknowledgments

This work was in part supported by Natural Sciences and Engineering Research Council of Canada Discovery Grant. The authors thank anonymous reviewers for helpful comments.

[10] http://scholar.google.ca/scholar?hl=en&as_sdt=1,5&as_vis=1&q=%22spinal+muscular+atrophy%22&scisbd=1, accessed Aug 13, 2013.

[11] All languages.

[12] http://www.ncbi.nlm.nih.gov/pubmed, accessed Aug 13, 2013.

[13] English only.

References

E. Balka and A. Butt. 2006. Information Seeking Practices for Youth, Parents and Seniors. *Report.* www.sfu.ca/act4hlth/

Bos, J. 2011. A Survey of Computational Semantics: Representation, Inference and Knowledge in Wide-Coverage Text Understanding. *Language and Linguistics Compass*, 5: 336–366.

S. Boytcheva, G. Angelova, I. Nikolova. 2012. Automatic Analysis of Patient History Episodes in Bulgarian Hospital Discharge Letters. *Proceedings of EACL*, 77-81

D. Brashers. 2001. Communication and Uncertainty Management. *Journal of Communication*, 51(3):477-497

D. Brashers, J. Neidig, S. Haas, L. Dobbs, L. Cardillo, J. Russell. 2000. Communication in the management of uncertainty: The case of persons living with HIV or AIDS. *Commun Monogr* , 67(1):63-84

Y. Cao, F. Liu, P. Simpson, L. Antieau, A. Bennett, J. Cimino, J. Ely, H. Yu. 2011. AskHERMES: An online question answering system for complex clinical questions, *Journal of Biomedical Informatics*, 44(2): 277-288.

K. Cohen, K. Verspoor, H. Johnson, C. Roeder, P. Ogren, W. Baumgartner Jr, E. White, H. Tipney, and L. Hunter. 2011. High-Precision Biological Event Extraction: Effects of System and Of Data. *Computational Intelligence*, 27: 681–701

M. Cornel, T. Rigter, S. Weinreich, P. Burgard, G. Hoffmann, M. Lindner, G. Loeber, K. Rupp, D. Taruscio and L. Vittozzi. 2013. A framework to start the debate on neonatal screening policies in the EU: an Expert Opinion Document, *European Journal of Human Genetics*, 1-6.

D. Demner-Fushman, J. Mork, S. Shooshan, A. Aronson. 2010. UMLS Content Views Appropriate for NLP Processing of the Biomedical Literature vs. Clinical Text. *Journal of Biomedical Informatics*, 43(4): 587–594.

Dunne, C., Shneiderman, B., Gove, R., Klavans, J. and Dorr, B. 2012. Rapid understanding of scientific paper collections: Integrating statistics, text analytics, and visualization. *J. Am. Soc. Inf. Sci.*, 63: 2351–2369

L. Emanuel and E. Emanuel. 1989. The medical directive: A new comprehensive advance care document. *Journal of the American Medical Association,* 261(22): 3288 – 3293.

G. Eysenbach. 2003. The impact of the Internet on cancer outcomes. *CA Cancer J Clin*, 53:356–71.

F. Farooq, F. Abadı´a-Molina, D. MacKenzie, J. Hadwen, F. Shamim, S. O'Reilly, M. Holcik, and A. MacKenzie. 2013. Celecoxib increases SMN and survival in a severe spinal muscular atrophy mouse model via p38 pathway activation. *Human Molecular Genetics*, 1–10.

S. Fox. 2011. *The Social Life of Health Information.* Pew Research Center's Internet & American Life Project,http://pewinternet.org/Reports/2011/Social-Life-of-Health-Info.aspx

S. Fox. 2011a. *Survey Questions.* Pew Research Center's Internet & American Life Project, http://pewinternet.org/Reports/2011/Social-Life-of-Health-Info.aspx

A. Henriksson and S. Velupillai. 2010. Levels of certainty in knowledge-intensive corpora: An initial annotation study. *Proceedings of the Workshop on Negation and Speculation in Natural Language Processing*, pages 41–45

J. Holmes. 1982. Expressing Doubt and Certainty in English. *RELC Journal*, 13:9-28.

L. Horn. 1989. *A Natural History of Negation*. The University of Chicago Press.

K. Hyland. 2006. Medical discourse: hedges. *Encyclopedia of Language and Linguistics,* p.p. 694- 697.

M. Liakata. 2010. Zones of conceptualisation in scientific papers: a window to negative and speculative statements. *Proceedings of the Workshop on Negation and Speculation in NLP*, 1–4.

W. McCoy, C. Alm, C. Calvelli, J. Pelz, P. Shi, A. Haake. 2012. Linking Uncertainty in Physicians' Narratives to Diagnostic Correctness. *Proceedings of the ACL-2012 Workshop on Extra-Propositional Aspects of Meaning in Computational Linguistics*

M. McMullan. 2006. Patients using the Internet to obtain health information: How this affects the patient–health professional relationship. *Patient Education and Counseling*, 63:24–28, Elsevier.

R. Morante and C. Sporleder. 2012. Modality and Negation: An Introduction to the Special Issue. *Computational Linguistics,* 38(2): 224-260.

R. Nawaz, P. Thompson, S. Ananiadou. 2010. Evaluating a Meta-Knowledge Annotation Scheme for Bio-Events, *Proceedings of the Workshop on Negation and Speculation in NLP*, pages 69–77.

D. Nicholas , P. Huntington, B. Gunter, C. Russell, R. Withey. 2003. The British and their use of the web for health information and advice: a survey. *Aslib Proc*, 55:261–76.

F. Oprescu, S. Campo, J. Lowe, J. Andsager, J. Morcuende. 2013. Online Information Exchanges for Parents of Children with a Rare Health Condition: Key Findings From an Online Support Community. *Journal of Medical Internet Research*, 15(1):e16

A.Raffle, M. Gray. 2007. *Screening. Evidence and Practice*. Oxford: Oxford University Press.

L. Sanchez, B. Li, C. Vogel. 2010. Exploiting CCG Structures with Tree Kernels for Speculation Detection. *Proceedings of CoNLL: Shared Task*, 126–131.

R. Saurı´ and J. Pustejovsky. 2009. FactBank: a corpus annotated with event factuality. Language Resources and Evaluation, 43:227–268.

G. Savova, W. Chapman, J. Zheng, R. Crowley. 2011. Anaphoric relations in the clinical narrative: corpus creation. *Journal of American Medical Informatics Association*, 18(4): 459-465.

D. Scott, R. Barone, B. Koeling. 2012. Corpus annotation as a scientific task. *Proceedings of LREC'2012*, p.p. 1481 – 1485.

M. Sokolova, K. El Emam, S. Chowdhury, E. Neri, S. Rose, E. Jonker. 2010. Evaluation of Rare Event Detection, *Advances in Artificial Intelligence 23*, pp. 379–383

M. Sokolova and G. Lapalme. 2011. Learing opinions in user-generated Web content", *Journal of Natural Language Engineering*, Cambridge University Press, 17(4): 541–567

S. Velupillai. 2010. Towards A Better Understanding of Uncertainties and Speculations in Swedish Clinical Text – Analysis of an Initial Annotation Trial. *Proceedings of the Workshop on Negation and Speculation in NLP*, 14–22.

V. Vincze, G. Szarvas, R. Farkas, G. Mora, and J. Csirik. 2008. The BioScope Corpus: Biomedical Texts Annotated for Uncertainty, Negation and their Scopes. *BMC Bioinformatics*, 9(Suppl 11):S9.

D. Wimalasuriya and D. Dou. 2010. Ontology-based information extraction: An introduction and a survey of current approaches. *Journal of Information Science*, 36(3):306 – 323

Author Index